OECD Public Governance Reviews

Public Procurement for Innovation

GOOD PRACTICES AND STRATEGIES

OECD
BETTER POLICIES FOR BETTER LIVES

This document was approved by the Public Governance Committee on 19 May 2017 and prepared for publication by the OECD Secretariat.

This document and any map included herein are without prejudice to the status of or sovereignty over any territory, to the delimitation of international frontiers and boundaries and to the name of any territory, city or area.

Please cite this publication as:
OECD (2017), *Public Procurement for Innovation: Good Practices and Strategies*, OECD Public Governance Reviews, OECD Publishing, Paris.
http://dx.doi.org/10.1787/9789264265820-en

ISBN 978-92-64-26581-3 (print)
ISBN 978-92-64-26582-0 (PDF)

Series: OECD Public Governance Reviews
ISSN 2219-0406 (print)
ISSN 2219-0414 (online)

The statistical data for Israel are supplied by and under the responsibility of the relevant Israeli authorities. The use of such data by the OECD is without prejudice to the status of the Golan Heights, East Jerusalem and Israeli settlements in the West Bank under the terms of international law.

Photo credits: Cover © nimon/Shutterstock.com; © Maciej Bledowski/Shutterstock.com; © Andrey_l/ Shutterstock.com. © Ismael Ruiz/Thenounproject.com; © Carla Dias/Thenounproject.com.

Corrigenda to OECD publications may be found on line at: *www.oecd.org/about/publishing/corrigenda.htm*.

Foreword

Governments are increasingly recognising the immense power of public procurement to solve global societal challenges, improve productivity and boost innovation, while ensuring value for money. Public procurement represents 12% of gross domestic product (GDP) and 29% of total government expenditures on average across OECD countries, a clear sign of its potential to support broader policy objectives, including the fostering of innovation.

Alongside sustainability and inclusiveness, innovation, which underpins jobs, productivity and growth, remains a policy imperative for all countries. This is why it was included in the Sustainable Development Goals as part of the universal 2030 Agenda for Sustainable Development.

The report presents insights from the OECD Survey on Strategic Innovation Procurement, covering 35 countries, and takes stock of the strategic use of procurement for innovation in OECD Member countries and non-Member economies. It provides evidence that public procurement is a major pillar of strategic governance and service delivery. Results from the survey show that governments are gradually implementing demand-side driven procurement policies to ensure that government-funded innovation results in better value for money in the future. Ground-breaking solutions that have resulted from these policies include liquid light-emitting diodes (LEDs), electric cars and robotic bed-washing facilities in hospitals. Such innovations are helping reduce energy consumption and support the transition to a low-carbon world.

The challenges of implementing an adequate public procurement strategy for innovation are many, such as reducing risk aversion, setting up new forms of co-ordination, improving skills and capacity, encouraging public purchasers to dialogue with suppliers, and enhancing data collection and the monitoring of results. To support countries in using public procurement to enhance innovation, the report provides evidence collected from the Survey and based on the 2015 OECD Recommendation of the Council on Public Procurement.

The findings presented in this report reinforce the OECD's vision for a holistic use of public procurement, and reaffirm the value of investing in this key tool both as a strategic function and as a means of addressing fundamental societal challenges. Indeed, through a more strategic use of procurement for innovation, countries can demonstrate quality governance worthy of citizens' and suppliers' trust and engagement, while delivering innovative solutions for servicing the public.

I invite countries to make use of this framework and learn from the good practice cases included in this report. Public procurement can and should be an integral part of our collective effort to support innovative and inclusive societies, in the OECD and beyond.

Angel Gurría
OECD Secretary-General

Acknowledgments

This report was prepared by the OECD Directorate for Public Governance and Territorial Development as part of its mission to help governments at all levels design and implement strategic, evidence-based and innovative policies to strengthen public governance.

Under the direction of Rolf Alter and guided by János Bertók, this report was led by Paulo Magina, co-ordinated and drafted by Anne Müngersdorff and received major chapter contributions by Lena Diesing. Further input was provided by Matthieu Cahen, Kenza Khachani and Jennifer Moreau. Survey and evaluation support was provided by Zsuzsanna Lonti, Guillaume Lafortune, Samuel Pinto Ribeiro and Minjoo Son; contributions on the Observatory Public Sector Innovation Survey were provided by Edwin Lau and Marco Daglio. Editorial assistance was provided by Pauline Alexandrov, Thibaut Gigou, Alpha Zambou, Julie Harris, Laura McDonald and Kjersti Berg.

It also benefited from valuable comments by Silvia Appelt, Maria Bravo, Mario Cervantes and Fernando Galindo-Rueda, as well for the OECD Survey by Michael Keenan, Sandrine Kergroach and Alistair Nolan of the OECD Directorate for Science, Technology and Innovation.

The OECD also wishes to express its gratitude for its fruitful co-operation with the European Research Area and Innovation Committee (ERAC), in particular to Minna Immonen and Frits Smulder of the ERAC-Secretariat, the ERAC Task Force on Innovation Procurement represented by Eva Buchinger, Austrian Institute of Technology GmbH, and Mart Laatsit, Ministry of Economic Affairs and Communications Estonia and the Public Governance Committee's Working Party of Leading Practitioners on Public Procurement (LPP Working Group).

Special thanks go to the ERAC and LPP Delegates and corresponding lead respondents from ministries and other organisations having responded to the OECD Survey on Strategic Procurement for innovation 2015: David Adolphe (DGE, France); Olga Anchishkina (NAPE, Russian Federation); Yasemin Aslan (TUBITAK, Turkey); Christos Aspris (Directorate General for European Programmes, Coordination and Development, Cyprus); Christopher Baird (Build in Canada Innovation Program Office of Small and Medium Enterprises Public Works and Government Services Canada, Canada); Jacqueline Barbara (Malta Council for Science and Technology, Malta); Stuart Barthropp (Department for Business Innovation and Skills, United Kingdom); Eva Buchinger (Austrian Institute of Technology AIT, Austria); Martin Buncek (Technology Agency of the Czech Republic, Czech Republic); Mario Calderini (Ministry of Research and Innovation, Italy); Myriam Cevallos (State Secreteriat for Education, Research and Innovation, Switzerland); Wolfgang Crasemann (BMWi, Germany); Floris den Boer (PIANOo, Netherlands); Johan Englund (Difi – Agency for Public Management and eGovernment, Norway); Juan Manuel Garrido (Ministry of Economy and Competitiveness, Spain); Shayne Gray (Ministry of Business, Innovation and

Employment, New Zealand); Cristina Guedes (General Directorate of Economic Activities, Portugal); Jussi Kajala (Tekes, Finland); Roman Kałamarz (Gmina Jarosław/Jarosław Commune, Poland); Aurelija Kazlauskienė (Ministry of Economy, Lithuania); Ljiljana Krejovic (Inovation Fund, Serbia); Trygve Olavson Laake (Difi – Agency for Public Management and eGovernment, Norway); Dávid Lakatos (National Development, Research and Innovation Office, Hungary); Lukas Levak (Ministry of Education, Youth and Sports, Czech Republic); Eivind Lorentzen (Norwegian Ministry of Trade, Industry / Department of Research and Innovation, Norway); Molina Maria de los Angeles (Dirección de Compras y Contratación Pública, Chile); Robbie Mckechnie (Inland Revenue, New Zealand); Pernille Nielsen (Danish Business Authorities, Denmark); Miguel Ortiz (CDTI, Spain); Christian Pecharda (FFG, Austria); Justyna Pożarowska (Public Procurement Office, Poland); Lucia Prozbikova (Ministry of Economy of the Slovak Republic, Slovak Republic); Jaime Quesado (eSPap, IP Portugal); Sigrid Rajalo (Ministry of Economic Affairs and Communications, Estonia); Angela Russo (Consip SpA, Italy); Karl Ryan (Office of Government Procurement, Ireland); Sara Hernández Muñoz (Secretaria de Economia, Mexico); Doris Scheffler (ZENIT GmbH, Germany); Kang-il Seo (Public Procurement Service, The Republic of Korea); Lisa Silvemark (Swedish Transport Administration, Sweden); Griet Storr-Hansen (Danish Agency for Science, Technology and Innovation, Denmark); Lena Svendsen (Ministry of Enterprise and Innovation, Sweden); Paul Thomson (Policy, Risk, Integrity and Strategic Management Sector - Public Services and Procurement Canada, Canada); Konstantinos Tzanetopoulos (General Secretariat of Commerce and Consumer Protection, Greece); Juergen Unger (Federal Procurement Agency, Austria); Waldo Van den Broeck (Diensthoofd Overheidsopdrachten Federale Overheidsdienst Personeel en Organisatie, Belgium); Christophe Veys (IWT since 2016 VLAIO, Belgium); Kirsti Vilén (Ministry of Employment and the Conomy, Finland); Vunnaporn Devahastin (Office of the Public Sector Development Commission, Thailand); Julia Wise (OMB, United States); Maria Margarita Zuleta Gonzalez (Colombia Compra Eficiente, Colombia).

This report was discussed and approved by the ERAC and LPP Delegates, and presented at the Forum on Procurement for Innovation on 5 October 2016 at OECD headquarters in Paris.

Table of Contents

Tables

Figures

Boxes

Acronyms and abbreviations

CCI	Chamber of Commerce and Industry
CPB	Central Purchasing Bodies
EC	European Commission
ERAC	European Research Area Innovation Committee
EU	European Union
GDP	Gross Domestic Product
IPR	Intellectual Property Rights
ISO	International Standards Organisation
IT	Information Technology
LPP	Working group of the Leading Practitioners on Public Procurement
NCPs	National Contact Points
OPSI	Observatory of Public Sector Innovation / OECD Internet Platform
PCP	Pre-Commercial Procurement
PPI	Public Procurement of Innovative solutions
R&D	Research and Development
SBIR	Small Business Innovation Research program
SBRI	Small Business Research Initiative programme
SDGs	Sustainable Development Goals
SMBA	Small and Medium Business Administration programme
SME	Small and Medium-sized Enterprise
SOE	State Owned Enterprise
UN	United Nations
UNCITRAL	United Nations Commission on International Trade Law
UNGA	United Nations General Assembly
WTO	World Trade Organization

Executive summary

Achieving the best result for the best price, public procurement can also be used by governments as a strategic instrument to promote innovation, achieve socio-economic and environmental policy objectives and address societal challenges.

This report presents existing government procurement strategies and practices for innovation and highlights the conditions needed for their implementation. Based on the evidence, it offers a framework for action that countries can use to support the strategic use of public procurement for innovation.

Key findings

A majority of countries support procurement for innovation

Almost 80% of responding countries support procurement for innovation, and 50% have developed an action plan for procurement for innovation, either as part of broader innovation or procurement strategies or as stand-alone initiatives.

Countries use various measures to support procurement for innovation, mostly policy instruments, regulations or legal instruments. Others include comprehensive programmes, e.g. on smart procurement in general or on research and development (R&D), followed by financial instruments, such as finance dedicated to procurement for innovation.

Countries pursue procurement for innovation to meet needs and demands for new products, goods or services

In most cases, demand for a new product or service was the main reason for using public procurement for innovation. The second most common reason cited by countries was to improve the performance of existing products or services achieving reduced costs and/or greater energy efficiency. Nearly half of the countries (49%) reported improved effectiveness or increased user satisfaction following the use of procurement for innovation. Experience shows how to combine successfully procurement for innovation with other policy areas, in particular to support small and medium-sized enterprises (SMEs) and promote environmentally sustainable solutions.

Partners in procurement for innovation practices

Successful procurement for innovation were carried out in collaboration with external partners, with the private sector representing 33%, followed by public institutions or bodies (27%), and research institutions (24%).

Main beneficiaries of procurement for innovation are citizens and the public service

Procurement for innovation benefitted mostly citizens and the public service as well as specialised groups such as patients or firefighters.

Main challenges faced

Countries have had to overcome a range of hurdles to implement their innovative procurement practices. The most common challenges were related to risk aversion, management, personnel and skills, capacity and political support. Sound measurement systems require robust data and indicators, and are crucial for evaluating innovation-procurement strategies and improving the return on investment as well as its impacts.

Successful strategic procurement for innovation

Successful strategic procurement for innovation requires governments to:

- communicate on the positive outcomes of innovation
- co-ordinate more closely in the horizontal and vertical management of tasks in governments
- demonstrate political leadership and political commitment
- build up the capacity, and numbers of skilled staff
- cultivate a more open culture towards new ways of working
- encourage co-operation between different branches of the public procurement process.

Meeting these objectives, however, is a challenging and long-term process.

Key areas for action

The OECD has developed a framework to support countries in their use of public procurement for innovation. The framework is adaptable to different country contexts, referring to new ways of competitive collaboration and new forms of networking between governmental and non-governmental actors. It includes **nine action areas** that should be present in any sound procurement for innovation agenda:

- Embed **policy strategies with defined targets** within any national, sub-national and regional innovation policy. This aims to secure strong political commitment. To achieve innovation as a secondary policy objective, public procurement must be deployed strategically in co-ordination with other policy areas.
- Set up a **legal framework,** including understandable definitions, guidelines and templates to facilitate its implementation.
- Designate "transformational" leaders with specialised knowledge to create skilled multidisciplinary teams, so as to encourage sound **management**. In addition, "intermediaries", e.g. an innovation agency, could help to bring together buyers and suppliers.

- Dedicate sufficient budgets, funds and other financial incentives, as lack of **financial support** is one of the main challenges in procurement for innovation.
- Promote **professionalisation** by providing specific training to build staff capabilities and skills, setting up multidisciplinary teams and competence centres focused on public procurement for innovation.
- **Raise awareness** by publishing good practice cases, creating a dedicated knowledge-sharing platform and/or hosting workshops and seminars to share and build success. Early **stakeholder engagement** should also not be underestimated.
- Undertake **risk management** and **measure impact** to reduce possible loss and damage, and increase trust.
- Define test standards, methods and quality certificates, using **standardisation** as a catalyst for innovation.
- Use appropriate **e-procurement** and **information technology (IT) tools** to carry out a proper risk assessment to measure impact.

Chapter 1

Public procurement for innovation: An overview

This chapter provides the general background for, and overview of, Public Procurement for Innovation: Good Practices and Strategies. *This includes a definition of procurement for innovation for the purposes of this report, an introduction to the methodology behind the findings, and an overview of the findings from the underlying OECD Survey on Strategic Procurement for innovation 2015.*

Note by Turkey: The information in this document with reference to "Cyprus" relates to the southern part of the Island. There is no single authority representing both Turkish and Greek Cypriot people on the Island. Turkey recognises the Turkish Republic of Northern Cyprus (TRNC). Until a lasting and equitable solution is found within the context of the United Nations, Turkey shall preserve its position concerning the "Cyprus issue".

Note by all the European Union Member States of the OECD and the European Union: The Republic of Cyprus is recognised by all members of the United Nations with the exception of Turkey. The information in this document relates to the area under the effective control of the Government of the Republic of Cyprus.

Public procurement represented 12% of gross domestic product (GDP) and 29% of total government expenditures on average across OECD countries (OECD, 2015a). In the European Union, it accounts for approximately 14% of GDP (European Commission, 2016). These figures illustrate the large budget size of public procurement and its significance for the economy as a whole and also indicate how the public sector can support policy objectives through public procurement.

In the field of innovation policies, governments have traditionally directed their efforts towards the supply side, ensuring that the private sector operates in an environment conducive to innovation. In recent years, however, the role of "demand-side policies" to support innovation has gained in prominence and has been receiving growing interest from many countries. Governments recognise that innovations materialise when there is a demand for innovation. Therefore, effective policies to support innovation have to focus on both supply and demand conditions (OECD, 2011).

Among demand-side innovation policies, such as innovation-friendly regulations or lead market initiatives, public procurement is increasingly recognised as a potential strategic instrument and a policy lever for achieving government policy goals, such as innovation, the development of small and medium-sized enterprises (SMEs), sustainable green growth and social objectives like public health and greater inclusiveness. All of these goals are in line with the aims specified in the 2030 Agenda for Sustainable Development and the related Sustainable Development Goals (OECD, 2016).

Public procurement for innovation has the potential to improve productivity and inclusiveness, if used strategically as targeted, demand-side innovation policies to meet societal needs. For example, it can anticipate future investments to address existing or future societal challenges; or it may allow potential vendors to enter the market with new, innovative goods or services, thus encouraging innovative solutions to pressing challenges.

The strategic use of public procurement to boost innovation is closely connected to a government's power to shape and create market conditions. In fact, given the size of public procurement, governments, among other actors, can influence demand on national or sub-national levels. Combining this influence with sectoral strategies can be useful to achieve targets in the above-mentioned critical areas.

As a result, the role of the purchaser in the public sector is changing to include more elements of active risk and benefit management. In the same way, to reap the benefits of procurement for innovation, the envisioned policy changes have to be well planned. Enhancing the implementation of strategic use of procurement for innovation requires strong political commitment, strategic management, capabilities to manage new organisational processes (Valovirta, 2015; Ongaro, 2015) and new ways of working across all levels of government (OECD, 2015b).

If governments want to address today's economic and societal challenges effectively, they need to demonstrate their capacity to deliver across all stages of the policy cycle with well-coordinated institutions and efficient alignment between long-term visions and short-term actions, and between budgetary decisions and regulatory instruments (OECD, 2015c). Moreover, they need to balance different policies and instruments productively. Part of governments' new role is also to facilitate interactions and collaborations between different actors. The flow of experiences, knowledge and skills is a prerequisite for innovation. This will demonstrate quality governance worthy of citizens' and suppliers' trust and engagement, while delivering innovative solutions for servicing the public.

The strategic use of public procurement to encourage innovation and tackle challenges works equally on national and sub-national levels. Across OECD countries, the near 134 000 sub-national governments are responsible for around 63% of public procurement, 59% of public investment and 40% of total government expenditure, as measured for the last OECD Public Governance Ministerial Meeting in Helsinki, Finland (OECD, 2015d). However, fragmentation of public demand on local regional and national levels can limit the pull of demand as responsibilities and co-operation between governments and agencies responsible for public procurement are often not co-ordinated with agencies and ministries in charge of innovation policies.

The OECD Recommendation of the Council on Public Procurement (OECD, 2015e) (hereafter referred to as the "OECD Recommendation") is a major step in substantiating the goal of strategic use of public procurement, including innovation. The recommendation develops the concept of balanced use of secondary policy objectives against primary procurement objectives (delivering goods and services necessary to accomplish government missions in a timely, economical and efficient manner). Box 1.1 details this principle as developed in the OECD Recommendation.

Following this, balanced does not necessarily refer to rank rather more as caveat, to understand the implications of policy choices and the need to consider what the prioritised objectives for the specific procurement process are. As an example, one of the principles in the EU procurement directives is to ensure that in the performance of public contracts economic operators comply with applicable obligations in the fields of environmental, social and labour law established by Union law, national law, collective agreements or by international environmental, social and labour law provisions.

Box 1.1. OECD Recommendation on Public Procurement: The principle of "Balance"

"V. RECOMMENDS that Adherents recognise that any use of the public procurement system to pursue secondary policy objectives should be balanced against the primary procurement objective.

To this end, Adherents should:

i) Evaluate the use of public procurement as one method of pursuing secondary policy objectives in accordance with clear national priorities, balancing the potential benefits against the need to achieve value for money. Both the capacity of the procurement workforce to support secondary policy objectives and the burden associated with monitoring progress in promoting such objectives should be considered.

ii) Develop an appropriate strategy for the integration of secondary policy objectives in public procurement systems. For secondary policy objectives that will be supported by public procurement, appropriate planning, baseline analysis, risk assessment and target outcomes should be established as the basis for the development of action plans or guidelines for implementation.

Box 1.1. **OECD Recommendation on Public Procurement: The principle of "Balance"** *(continued)*

iii) Employ appropriate impact assessment methodology to measure the effectiveness of procurement in achieving secondary policy objectives. The results of any use of the public procurement system to support secondary policy objectives should be measured according to appropriate milestones to provide policy makers with necessary information regarding the benefits and costs of such use. Effectiveness should be measured both at the level of individual procurements, and against policy objective target outcomes. Additionally, the aggregate effect of pursuing secondary policy objectives on the public procurement system should be periodically assessed to address potential objective overload."

Source: OECD (2015e), "Recommendation of the Council on Public Procurement", www.oecd.org/gov/ethics/OECD-Recommendation-on-Public-Procurement.pdf

Defining strategic use of public procurement for innovation

For the purposes of this report, the strategic use of public procurement for innovation is defined as **any kind of public procurement practice (pre-commercial or commercial) that is intended to stimulate innovation through research and development and the market uptake of innovative products and services**. This definition follows the approach of the European Research Area and Innovation Committee (European Council, 2015).

Throughout literature, different expressions, sometimes with diverging scope, e.g. excluding pre-commercial procurement (Edquist et al., 2015; Edquist and Zabala-Iturriagagoitia, 2012), cover what this report labels as "strategic use of public procurement for innovation".

Concerning the modernisation of the procurement process, this report includes only procurement for innovation related aspects (e.g. risk management, life-cycle cost and e-procurement) and focusses mainly on the procurement of research and development, or procurement of an innovative product.

Background and methodology

This report is based on responses to the OECD Survey on Strategic Procurement for innovation 2015, (hereafter referred to as the "OECD Survey"), which was carried out at the end of 2015 (see Annex C). The work originates from the call of the OECD Working Group of the Leading Practitioners in Public Procurement (LPP) in April 2015 to design a report on Strategic Procurement for innovation, gathering evidence on the state of play, collecting innovation good practices and providing further guidance on the strategic use of public procurement. The OECD Survey is a contribution of the Public Governance Committee to the OECD Inclusive Growth Strategy, the OECD Innovation Strategy and the OECD Green Growth Strategy. It is endorsed by the LPP Working Group and the ERAC Secretariat.

The OECD Survey was structured into two parts and sent to LPP Delegates in OECD countries, ERAC Delegates in the EU member states, as well as to OECD accession/partnership and associated countries and economies. Answers were provided by the LPP and ERAC Delegates as a consolidated response (one consolidated answer per country).

Part I of the OECD Survey was policy oriented and based on the first six questions of the ERAC Questionnaire on Procurement for innovation 2014 (ERAC, 2015), with a focus on the strategic dimension, implementation and impact at the national level. This part was developed in collaboration with the ERAC Secretariat.

Part II was based on the OECD Observatory Public Sector Innovation Survey (OPSI) (OECD, 2015f), with the purpose of collecting good practice cases on strategic procurement for innovation in national and sub-national contexts, to be published on the OPSI platform.

Overview of the responses to the OECD Survey on Strategic Procurement for innovation 2015

In total, 35 OECD member countries[1] participated in the OECD Survey and 80% (28 OECD member countries) responded: Austria, Belgium, Canada, Chile, Czech Republic, Denmark, Estonia, Finland, France, Germany, Greece, Hungary, Ireland, Italy, Korea, Mexico, Netherlands, New Zealand, Norway, Poland, Portugal, Slovak Republic, Spain, Sweden, Switzerland, Turkey, United Kingdom and the United States. In addition, the following seven OECD non-member countries participated: Colombia, Cyprus,[2] Lithuania, Malta, Russian Federation, Serbia and Thailand.

Of the 35 responding countries, 34 provided information on the state of play of policy strategies. Of the total responding countries, 30 (88%) provided good practice cases of procurement for innovation, and 23 (68%) released them for publication (see Tables 1.1 and 1.2). The good practice cases are presented throughout this report in boxes as "spotlights".[3]

The OECD Survey took stock of definitions of public procurement for innovation that countries use for policy purposes. Scope and definitions for procurement for innovation policies in countries vary. European Union countries were obliged to transpose the new 2014 EU Public Procurement Directives (European Commission, 2014) into national law, which covers opportunities related to procurement for innovation, e.g. the innovation partnership, the exemption on Pre-Commercial Procurement (PCP) and facilitating Public Procurement of Innovative solutions (PPI).

Some countries, including Ireland, the Netherlands and the United Kingdom, also pursue a modified model of the Small Business Innovation Research program used in the United States ("SBIR type") (SBIR, 2016). This programme promotes the participation of small businesses in federal research and development opportunities. The multiplicity of definitions is reflected in the survey responses where nine of the responding countries have developed their own definitions; six countries have no definitions and one country follows the rules in line with the World Bank Procurement Guidelines.

While the policy-related questions were answered throughout by national organisations, the good practices cases were provided by different public institutions on national, regional and local levels, including Central Purchasing Bodies (CPBs), agencies in charge of public procurement, ministries, and other entities tasked with promoting businesses or innovation. Overall, 87% of these respondents are situated at the national level; 10% at sub-national and 3% at the local level. Figure 1.1 presents the main sectors of the organisations that submitted good practice cases. Most good practices came from organisations in charge of general public services, a category that includes functions related to overall governance, co-ordination between different levels, foreign affairs, and general research capabilities. The second major category represents organisations in charge of economic affairs, followed by health- and education-related institutions.

Figure 1.1. **Sectors of the organisations providing examples of good practices**

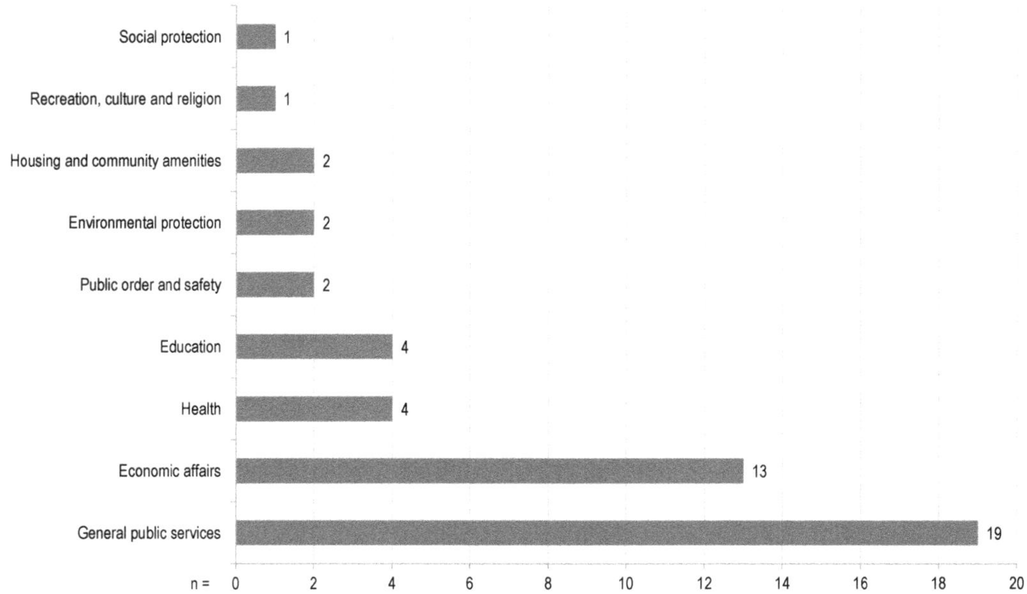

Note: Countries could provide multiple responses; n= numbers of responses provided.

Source: Country responses to OECD (2017), "OECD Survey on Strategic Procurement for innovation 2015", in *Public Procurement for Innovation: Good Practices and Strategies*, Annex C, OECD Publishing, Paris.

The most relevant areas mentioned in the good practice cases are presented in Chapter 2: how procurement for innovation was triggered, how partners supported the process, and in terms of impact assessment, what the expectations and results of the innovative practice were.

The good practice cases fall broadly into one of two categories: 1) projects; and 2) programmes or initiatives (see Tables 1.1 and 1.2). Countries were asked to describe their submission with up to five keywords to capture the key elements of the innovation.

Table 1.1. **Good example of practices of strategic procurement for innovation**

Country	Project	Subject	Keywords chosen by countries
Austria[1]	Full range (socially) sustainable food package with dynamic allergen indication	Food database	Food; Multi-quality-strategy; Dynamic allergen information tool
	MOVEBAG and MOVEBEST - Mobile traffic management system for roadworks and major incidents	Traffic management	Mobile traffic management; Mobile traffic detection
Belgium[2]	Smart@Fire - Smart Personal Protective Systems for fire fighters (Region of Flanders)	Smart Personal Protective System	ICT - Localisation systems; Intelligent personnel protective systems; Integrated systems
Czech Republic	Archive management for State Archives - in line with the BETA programme	Software module	Archive services; Archive process; Hard-copy documents; Safe communication; Document deposition place management
Finland	iLOQ - Energy-efficient locks (Oulu, Jyväskylä and Kuopio Regions)	Digital locks Energy	Digital locks; Energy-efficient locks; Locking system; Energy-efficiency; Digitalisation
France	Liquid LEDs (bulb for public lighting)	Liquid LED technology	Total cost of ownership; Energy savings; More functionalities (video monitoring); Efficiency; Quality
Germany	THALEA - Improve care for acutely life-threatened patients by telemedicine and telemonitoring (Region of North-Rhine Westphalia)	Telemedicine	Telemedicine; Interoperability of Patient Data Management; Systems Decision-support; Closing innovation gap; Sharing expertise
Hungary[3]	Smart@Fire - Smart Personal Protective Systems for fire fighters (Észak-Alföld Region)	Smart Personal Protective System	PPS; PPE; firefighter gear; ICT; PCP
Netherlands	Robotic Bed-washing Facility in Hospitals (Erasmus Medical Centre, Erasmus University Rotterdam)	Robotic	Clean hospital beds; Lower total costs of ownership; Lower energy costs; Less detergents; CO_2 reduction
Norway	Omsorg + Kampen - Smart house platform for senior housing residence (City of Oslo)	Technology for elderly people	Increased efficiency through user-driven integration of welfare technology
Poland	Hybrid lighting in the Jarosław commune (City of Jarosław)	Lighting	Safety; Clean energy; Environmental protection
Sweden	Electrified Roads - knowledge base for industrial, academic and political decisions	Heavy traffic	Electrified roads; Heavy traffic; PCP; Triple-helix co-operation; Environment

Notes: All good practice projects presented above have been released to be published.

1. Austria submitted two released good practice cases.

2. Belgium provided a good practice case on Smart Personal Protective System, same case provided by Hungary.

3. Hungary provided a good practice case on Smart Personal Protective System, same case provided by Belgium.

Source: Country responses to OECD (2017), "OECD Survey on Strategic Procurement for innovation 2015", *in Public Procurement for Innovation: Good Practices and Strategies*, Annex C, OECD Publishing, Paris.

These projects have different approaches - set up as European joint cross-border projects or on national and sub-national levels; seven of these projects are managed on regional or local levels. This presentation of good practice examples underlines the fact that most innovative projects are related to new information and communication technologies.

Table 1.2 provides an overview of programmes or initiatives presented as good practice cases. These programmes were implemented on national level and aim to achieve a range of objectives, such as energy efficiency and smarter use of information and communication technology.

Table 1.2. **Programmes or initiatives submitted as good practice examples of strategic procurement for innovation**

Country	Programme/initiative	Keywords chosen by countries
Canada	The Build in Canada Innovation Program (BCIP) contracts to entrepreneurs with pre-commercial innovations	Innovative; Competitive; Engagement; Supply(technology)-push; Collaboration
Chile	Workshop to Improve Regional Public Procurement (Public procurement management of municipalities)	Problem definition; Listen to the user's opinions; Multidisciplinary work
Colombia	The National Development Plan institutes procurement for innovation as a cross-cutting public policy	MinTic: Lab; Test-scenarios; Cybersecurity; Information; IT-management ANSPE: Poverty; Education; Technology; Income; Social-innovation EPM: Water-loss; Leak; Costs; Consumer; Innovation
Ireland	Small Business Innovation Research (SBIR) - Electric Vehicle Smart Charging SME support	Solutions; Efficiencies; Savings; Job creation; Export opportunities
Italy	Integrated Energy Service Framework Agreement 3	Energy efficiency; Sustainability; Standardisation; Modernisation; Leading by example
Korea	Excellent Government Supply Products Program, Certification of excellent technology and quality	Technology certification; SME; Framework agreement; Direct online orders; Promotion
New Zealand	APP4IR Crowd Sourcing for start-ups	Crowd source; Prize; Dragons den
Russia	Contracts - Suppliers are paid for the final results that they achieved according to clear and measured indicators of the final effects	Faster; Easier; Cheaper; Better results than standard technology; Product, service
Spain	Public Procurement of Innovation Policy in Spain	Co-finance; Reference; Leverage; Jump and opportunity
Thailand	Integrity Risk Assessment in Public Procurement in Thailand	Integrity risk mitigation measures; Guideline for mitigating integrity risks; Public procurement in reform; A risk indicator system; Sound and modern public procurement law; Institutional capacity of the office of public procurement management
United Kingdom	SBRI supports development of long-endurance unmanned vessel for oceanographic research	New technologies; Cheaper; More efficient

Note: All good practice projects presented above have been released to be published.

Source: Country responses to OECD (2017), "OECD Survey on Strategic Procurement for innovation 2015", in *Public Procurement for Innovation: Good Practices and Strategies*, Annex C, OECD Publishing, Paris.

Framework to promote the strategic use of public procurement for innovation

This report takes stock of the state of play in procurement for innovation. Building on the lessons of the policies and practices, it also provides a framework to help countries implement strategic procurement for innovation in Chapter 3.

The framework is designed as a modular and flexible structure and can be applied in a variety of circumstances and levels of governance at national and sub-national levels and across sectors. The framework illustrates the fundamental requirements of good public governance to integrate the strategic innovative approach in public procurement. It highlights nine areas of action from an integrated perspective to improve co-ordination, governance, management and communication, among other measures required for success.

Notes

1. In 2015, all 34 OECD member countries were involved in the OECD Survey, and 28 of them responded. Since 1 July 2016, OECD has 35 member countries: Latvia deposited its instrument of accession to the OECD Convention on 1 July 2016, thereby becoming a full member of the Organisation. For more information about OECD members and partners, see www.oecd.org/about/membersandpartners.

2. *Note by Turkey*: The information in this document with reference to “Cyprus” relates to the southern part of the Island. There is no single authority representing both Turkish and Greek Cypriot people on the Island. Turkey recognises the Turkish Republic of Northern Cyprus (TRNC). Until a lasting and equitable solution is found within the context of the United Nations, Turkey shall preserve its position concerning the “Cyprus issue”.

   Note by all the European Union Member States of the OECD and the European Union: The Republic of Cyprus is recognised by all members of the United Nations with the exception of Turkey. The information in this document relates to the area under the effective control of the Government of the Republic of Cyprus.

3. For good practice cases on the state of play with regard to policy strategies (based on Part I of the OECD Survey), see Annex A with case studies by country in alphabetical order. For the state of play of good practice cases (based on Part II of the OECD Survey), see the “spotlights” boxes throughout this report.

References

Edquist, C. et al. (eds.) (2015), *Public Procurement for Innovation,* Edward Elgar Publishing, https://charlesedquist.com/2015/01/27/public-procurement-for-innovation/.

Edquist, C. and J.M. Zabala-Iturriagagoitia (2012), "Public procurement for innovation as mission oriented innovation", *Research Policy*, www.researchgate.net/publication/256921387_Public_Procurement_for_Innovation_as_mission-oriented_innovation_policy

ERAC (European Research Area and Innovation Committee) (2015), "ERAC Opinion on Innovation Procurement", based on "Questionnaire on Innovation Procurement 2014", http://data.consilium.europa.eu/doc/document/ST-1209-2015-INIT/en/pdf.

European Commission (2016), "Public procurement", http://ec.europa.eu/growth/single-market/public-procurement/index_en.htm.

European Commission (2014), "Procurement Directive 2014/24/EU, replacing Directive 2004/18/EC", http://eur-lex.europa.eu/legal-content/EN/TXT/?uri=celex:32014L0024 and "Procurement Directive 2014/25/EU, replacing Directive 2004/17/EC (Utilities)", http://eur-lex.europa.eu/legal-content/EN/TXT/?uri=celex:32014L0025.

European Council (2015), "European Research Area and Innovation Committee (ERAC)", www.consilium.europa.eu/en/council-eu/preparatory-bodies/european-research-area-innovation-committee/.

OECD (2017), "OECD Survey on Strategic Procurement for innovation 2015", in *Public Procurement for Innovation: Good Practices and Strategies*, Annex C, OECD Publishing, Paris.

OECD (2016), "Towards an OECD Strategic Response to the Sustainable Development Goals: Consultation draft", February, www.oecd.org/dac/Consultation-draft--Outline-OECD-strategic-response-to-the-SDGs.pdf.

OECD (2015a), *Government at a Glance 2015*, OECD Publishing, Paris, http://dx.doi.org/10.1787/gov_glance-2015-en.

OECD (2015b), *The Innovation Imperative in the Public Sector: Setting an Agenda for Action*, OECD Publishing, Paris, http://dx.doi.org/10.1787/9789264236561-en.

OECD (2015c), "Public governance for inclusive growth: Towards a new vision for the public sector", closing remarks, Mari Kiviniemi, Deputy Secretary-General, OECD, 28 October, www.oecd.org/governance/ministerial/mari-kiviniemi-closing-remarks.pdf.

OECD (2015d), "Policy Shaping and Policy Making: The Governance of Inclusive Growth", www.oecd.org/governance/ministerial/the-governance-of-inclusive-growth.pdf.

OECD (2015e), "Recommendation of the Council on Public Procurement", www.oecd.org/gov/ethics/OECD-Recommendation-on-Public-Procurement.pdf.

OECD (2015f), "Observatory of Public Sector Innovation Platform", www.oecd.org/governance/observatory-public-sector-innovation/.

OECD (2011), *Demand-side Innovation Policies*, OECD Publishing, Paris, http://dx.doi.org/10.1787/9789264098886-en.

Ongaro, E. (2015), *Multi-level Governance: The Missing Linkages,* www.emeraldinsight.com/doi/book/10.1108/S2045-794420154.

SBIR (Small Business Innovation Research) (2016), "About SBIR", www.sbir.gov/about/about-sbir.

Valovirta, V. (2015), "Building capability for public procurement of innovation", in Edquist et al. (eds.) *Public Procurement for Innovation*, Edward Elgar Publishing, https://charlesedquist.com/2015/01/27/public-procurement-for-innovation/.

Chapter 2

The state of play in strategic procurement for innovation

This chapter presents the state of play with regard to strategic use of procurement for innovation in OECD member countries and non-member economies. It presents the findings in six sections that lead up to the development of a framework to support procurement for innovation: 1) policies, strategies and instruments employed by countries to support procurement for innovation; 2) objectives of, and results following from, procurement for innovation as set out by the countries; 3) partners and beneficiaries of procurement for innovation; 4) challenges pertaining to the implementation of procurement for innovation; 5) lessons learned or successful levers to tackle these challenges; and 6) how different policy objectives can be combined.

Note by Turkey: The information in this document with reference to "Cyprus" relates to the southern part of the Island. There is no single authority representing both Turkish and Greek Cypriot people on the Island. Turkey recognises the Turkish Republic of Northern Cyprus (TRNC). Until a lasting and equitable solution is found within the context of the United Nations, Turkey shall preserve its position concerning the "Cyprus issue".

Note by all the European Union Member States of the OECD and the European Union: The Republic of Cyprus is recognised by all members of the United Nations with the exception of Turkey. The information in this document relates to the area under the effective control of the Government of the Republic of Cyprus.

The results from the OECD Survey on Strategic Procurement for innovation 2015 (hereafter, the "OECD Survey"), show that in many countries the use of procurement for innovation has been included in national or sub-national innovation strategies. Results also show that there is room for improvement in terms of the implementation of professional guidance, exchange of experiences and good practices, and the collection of reliable performance data. To fully exploit the potential of strategic procurement for innovation these activities should be part of the implementation process.

In any condition, the assessment of the situation confirms the existence of a complex set of challenges, mirroring the already known complexities of procurement for innovation.

Policies and instruments to support procurement for innovation

So far, countries have made progress in encouraging and developing procurement for innovation using a variety of policies, strategies and other instruments. Half of responding countries have developed an action plan that substantiates the strategy in concrete terms (see Figure 2.1).

Figure 2.1. **Use of action plans to boost procurement for innovation at the national level**

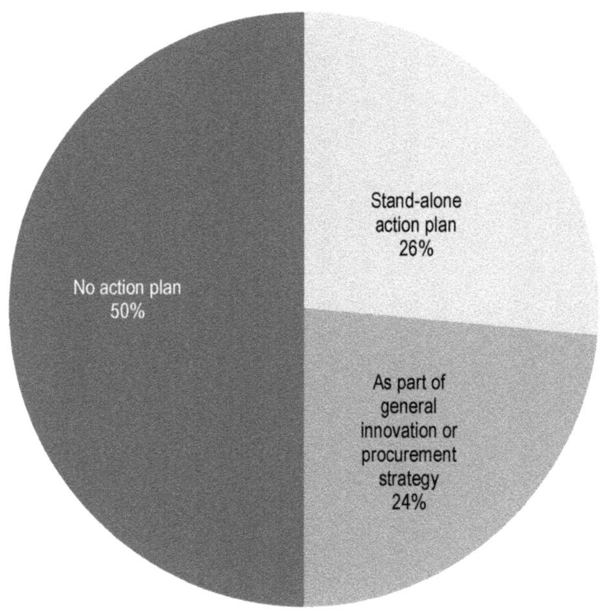

Source: Country responses to OECD (2017), "OECD Survey on Strategic Procurement for innovation 2015", in *Public Procurement for Innovation: Good Practices and Strategies*, Annex C, OECD Publishing, Paris.

Some 26% of responding countries developed a stand-alone action plan (see Table 2.1), sometimes focused on specific sectors. The other 24% of responding countries reported that the action plan is part of the country's general innovation or procurement strategy (see Table 2.2 as well).

Table 2.1. **Examples of countries with a stand-alone procurement for innovation action plan**

Austria	Austria has established the "Austrian Action Plan on Public Procurement Promoting Innovation PPPI" (2012) as a follow up of the "Austrian Strategy for Research, Technology and Innovation RTI (2011)". The RTI strategy aims to create a "systemic, modern policy on research, technology and innovation" by using public procurement as one of the levers. The PPPI action plan outlines in detail how this leverage effect will be achieved (i.e. measures, resources, responsibilities) and its implementation is progressing very well.
Canada	The federal government of Canada has an Economic Action Plan, part of which is the "Build in Canada Innovation Program (BCIP)". Canada's innovation strategy, entitled "Seizing Canada's Moment", is overseen by Innovation, Science and Economic Development Canada.
Denmark	The framework for procurement for innovation is part of a national procurement strategy. In October 2013 the Danish government launched a "Strategy for Intelligent Public Procurement".
France	The framework of procurement for innovation in France is part of the innovation strategy as a demand-side support tool. The main objective is to support the growth of innovative small and medium-sized enterprises (SMEs) by funding the development of their innovations, providing them with access to new markets and quality references.
Mexico	In 2013, President Enrique Peña Nieto instructed the Ministry of Economy to create a programme to drive innovation through public procurement.
Netherlands	In the Netherlands there is a stand-alone procurement for innovation action plan: "Innovatiegericht Inkopen".
Turkey	Turkey's "Program for Technology Development and Domestic Production through Public Procurement" is one of the 25 primary transformation programmes within the frame of 10th National Development Plan (2014-18).
United States	The United States has a stand-alone action plan on procurement for innovation, issued by the Office of Management and Budget in 2010, and titled "25 Point Implementation Plan to Reform Federal Information Technology Management".
Russian Federation	The Russian Federation specifies requirements related to procurement for innovation in the law, including obligations (as percentage shares) for innovation products to be procured. State-owned enterprises (SOEs) are obliged to purchase innovations and to publish their plans for procurement for innovation. Initially, the target is set at 2.5%; this target will be increased to 5%.

Source: Country responses to OECD (2017), "OECD Survey on Strategic Procurement for innovation 2015", in *Public Procurement for Innovation: Good Practices and Strategies*, Annex C, OECD Publishing, Paris.

Table 2.2 lists the countries that provided examples of an procurement for innovation action plan as part of the country's general innovation or procurement strategy.

Table 2.2. **Examples of countries with a general action plan as part of other strategies**

New Zealand	New Zealand has an enabling approach towards public procurement fostering innovation. The policy framework provides a flexible and supportive environment for procurement generating new and improved solutions.
Portugal	Portugal does not have a specific strategic framework for procurement for innovation or a stand-alone procurement for innovation action plan. Nevertheless, the general legal system in Portugal supports procurement for innovation and specifies the scope for procurement for innovation policy, Public Contracts Code (2008).
Spain	Spain's procurement for innovation action plan is both part of the country's general innovation strategy and part of the procurement strategy.
Sweden	Sweden does not have a specific procurement for innovation action plan. Instead, procurement for innovation is embedded into the Swedish Innovation Strategy (2012).
United Kingdom	The United Kingdom's main vehicle for taking forward procurement for innovation is the Small Business Research Initiative (SBRI).
Colombia	The Colombian National Development Plan (2014-18) specifies procurement innovation as a cross-cutting strategy targeted to generate a higher economic and social value to enhance the conditions for the development of business activities.

Table 2.2. **Examples of countries with a general action plan as part of other strategies** *(continued)*

Lithuania	Utmost attention at the moment in Lithuania is given to the development of demand-side measures. Even though the country does not have a specific procurement for innovation action plan, procurement for innovation is a part of Lithuanian Innovation Development Programme for 2014-20.
Malta	Malta has limited experience with procurement for innovation, and does not have a stand-alone procurement for innovation policy. However, existing procurement structures do allow for procurement for innovation to take place and a number of examples of the application of procurement for innovation exists.

Source: Country responses to OECD (2017), "OECD Survey on Strategic Procurement for innovation 2015", in *Public Procurement for Innovation: Good Practices and Strategies*, Annex C, OECD Publishing, Paris.

In addition, further countries provided examples of other policy initiatives for procurement for innovation (see Table 2.3). These examples reflect how support for procurement of innovation can also be found in other policy initiatives (such as policies related to a national knowledge base, entrepreneurship, overall national innovation framework, etc.)

Table 2.3. **Examples of other policy initiatives for procurement for innovation in OECD countries**

Estonia	The Ministry of Economic Affairs and Communications and the Ministry of Education and Research both developed strategies that clarify the concept of procurement for innovation and at the same time call on a need to further elaborate it. The strategies are entitled "Knowledge based Estonia 2014-2020" and "Estonian Entrepreneurship Growth Strategy 2014-2020".
Finland	There is no stand-alone procurement for innovation action plan in Finland; however, the country has an overall national strategic framework with objectives. The implementation takes place through various sectors and sector strategies. This allows for ownership and takes into account sector-specific characteristics and demands.
Germany	Procurement for innovation is part of the overall innovation strategy of the German federal government. The "High-Tech Strategy – Innovation for Germany" encompasses all research, technology and innovation measures of the German government. Innovative procurement is the most important measure under the framework of demand-oriented policy instruments. Six federal German ministries agreed in 2007 to promote innovation-oriented public procurement.
Korea	To promote public procurement of innovation and ensure SME access, Korea operates the New Technology Purchasing Assurance programme, which includes elements to encourage procurement-conditioned research and development (R&D) by SMEs.

Source: Country responses to OECD (2017), "OECD Survey on Strategic Procurement for innovation 2015", in *Public Procurement for Innovation: Good Practices and Strategies*, Annex C, OECD Publishing, Paris.

The absence of a procurement for innovation strategy or action plan does not preclude initiatives related to procurement for innovation. Even though half of the respondents do not have a dedicated procurement for innovation action plan at the national level, almost 80% of the respondent countries reported having taken at least one or more specific actions in support of procurement for innovation. For example, some countries (e.g. Chile, Czech Republic, Cyprus[1], and Serbia) reported good practice cases without confirming having a valid action plan. Those cases have been drafted as pilots or singular test cases, with the purpose of building expertise and trust in new procedures, or to see what kind of obstacles need to be overcome.

Finally, the European Union's policy strategy includes procurement as one of the levers to stimulate innovation (see Box 2.1).

Box 2.1. European Union policy perspective

To create a more innovative Europe, the European Union's "*Europe 2020 Strategy*" follows a strategic approach to bridge the innovation gap by using demand-side policies, e.g. through public procurement. This strategy is embedded in the European Union's flagship initiatives, e.g. "Innovation Union" or "A Digital Agenda for Europe".

Source: Adapted from European Commission (2016a), "Europe 2020", http://ec.europa.eu/europe2020/index_en.htm.

For the purpose of the OECD Survey, the following types of instruments in support of procurement for innovation are distinguished (see Figure 2.2):

1. policy instruments, such as overarching strategies, legal frameworks and targets, guidance, policy papers and similar documents
2. programmes, which are considered structured plans with the purpose of achieving a defined target for a certain policy area
3. financial instruments, which comprise monetary incentives dedicated to procurement for innovation or specialised funds to finance innovative practices
4. stand-alone cases, which are one-off procurement projects.

Some countries have mentioned more than one category.

Figure 2.2. **Type of instruments to support procurement for innovation**

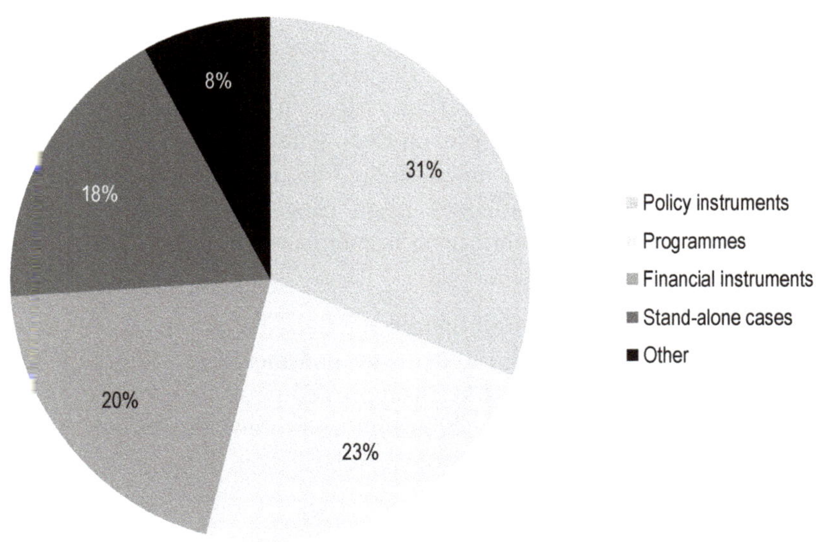

Note: Countries could provide multiple responses.

Source: Country responses to OECD (2017), "OECD Survey on Strategic Procurement for innovation 2015", in *Public Procurement for Innovation: Good Practices and Strategies*, Annex C, OECD Publishing, Paris.

Policy instruments are the most common type used to support procurement for innovation. These can take different forms. For example, as part of the legal framework, countries set targets for procurement for innovation, issue guidelines or regulations, or set high-level strategies in support of procurement for innovation.

Comprehensive programmes on the national level are the second most used instrument to support strategic procurement for innovation. However, not all the programmes mentioned are necessarily targeted at procurement for innovation alone. Most of the programmes have a different focus, for example on smart procurement in general, or on research and development (R&D), or are sector-related (e.g. energy), and include procurement for innovation as one factor within a broader topic.

Financial instruments are the third type of instrument used. In most cases, a specific pool of funds was dedicated to procurement for innovation.

The instruments that were assigned to the category "Other" included a range of different initiatives and are sometimes implemented in addition to an procurement for innovation action plan, e.g. New Zealand created a commercial pool of experts and the United States considered it useful to collect portfolio statistics.

Objectives and results from good practice cases

Many times, habits that have remained unchanged for decades have proven to be the largest obstacles countries have to overcome when aiming to increase innovative procurement practices, as procurement for innovation needs agile governance (OECD, 2015). An example of this is the tendency to use the award criteria of lowest price. To promote innovative solutions in procurement using the Most Economically Advantageous Tender (MEAT) criteria and use of a life-cycle cost (LCC) approach might work better.

Analysing the objectives that countries pursued and their expectations in doing so can shed light on how to create favourable conditions and an enabling environment for future innovations.

Countries provided different reasons why, in a specific case, they chose an innovation-oriented procurement approach over a traditional one, and what expectations they place on procurement for innovation. The reason for using procurement for innovation to meet societal challenges often relates to a concrete need or concrete demand. The reasons countries choose to implement procurement for innovation can be grouped into two categories:

1. Most of the respondents highlighted the need for goods or services that were not yet available to those with the demand, and therefore required a specialised, new good or service, as opposed to an improved good or service (see the example of the European PCP-Project Smart@Fire submitted twice, by Belgium and Hungary; or the United Kingdom's submission on a vessel for oceanographic research).
2. Another set of good practice cases presented in this report aimed at improving the performance of existing products or services, such as producing total cost savings and/or greater energy efficiency and risk reduction.

Boxes 2.2 and 2.3 feature "spotlights"[2] on Belgium, region of Flanders, and the United Kingdom, whose good practice submissions relate to filling concrete demands by implementing strategic procurement for innovation.

Box 2.2. Spotlight: Innovative practice in Belgium

European PCP Project Smart@Fire (2012)

Continuously operating in perilous situations, firefighters need a solution to monitor, measure, interpret and act on the environment. The solution must combine safety and comfort in all situations. The main scope of the project is to reduce firefighter injuries and casualties. Based on an in-depth needs assessment, Smart@Fire envisions the next generation Smart Personal Protective Systems (PPS).

Some 961 European fire and rescue services were consulted about their expectations for innovation for the smart PPS. The next step was to organise market consultations where technology suppliers and procurers engaged with each other and fine-tuned the scope of the prototype. These consultations were held in three different countries (Belgium, France and Germany) and showed that a smart PPS holds high potential for innovation from a technological perspective, summarised in the following challenges:

- PPS central nerve system covering: system architecture, communication, localisation, visualisation and interfaces (with temperature and explosive gas stand-alone devices)
- IR thermal hotspot detector
- HMD/HUD firefighter visualisation system
- "BE SEEN" omnidirectional active illumination.

For additional information, see www.smartatfire.eu.

Source: Country response to OECD (2017), "OECD Survey on Strategic Procurement for innovation 2015", in *Public Procurement for Innovation: Good Practices and Strategies*, Annex C, OECD Publishing, Paris.

The following spotlight highlights the use of an existing technology and a new sensor technology.

Box 2.3. Spotlight: Innovative practice in United Kingdom

SBRI supports the development of a long-endurance, unmanned marine vessel for oceanographic research (2014)

The SBRI competition invited proposals for long-endurance, marine, unmanned surface vehicles (LEMUSV) that could use both existing and new sensor technology to gather data from the oceans for several months at a time. ASV (a Portsmouth-based SME) used the initial GBP 50 000 funding from Phase 1 of the programme to develop the concept for the C-Enduro, a rugged, self-righting vehicle that uses solar panels, a wind generator and a lightweight diesel generator as energy sources to keep the vessel at sea for up to three months. The success of this concept led ASV to Phase 2 of the competition where they were awarded GBP 390 000 to build a prototype.

For more information, see www.gov.uk/government/case-studies/c-enduro-a-boat-that-goes-the-distance and www.gov.uk/government/news/c-enduro-world-wildlife-fund-research-project.

Source: Country response to OECD (2017), "OECD Survey on Strategic Procurement for innovation 2015", in *Public Procurement for Innovation: Good Practices and Strategies*, Annex C, OECD Publishing, Paris.

Countries cited additional reasons for pursuing strategic procurement for innovation. One is related to an overarching or higher level goal, such as increasing knowledge or awareness, or supporting innovative and small- and medium-sized firms as a stand-alone policy goal that was established irrespective of the policies pursued in public procurement. Some of the good practices cited by countries relate to the development of larger programmes, instead of procurement for a specific innovative good or service to solve a demand.

Another reason is related to a solution provided by a supplier. In one example, the supplier highlighted an innovative solution. The demand was not necessarily identified, given that the potential for different solutions to the problem was unknown. However, the supplier anticipated interest in a solution that provided a better result than the buyer expected and suggested the improved solution – with success (see Finland's spotlight in Box 2.4).

Box 2.4. Spotlight: Innovative practice in Finland

Digital, energy-efficient locks, iLOQ (2007)

When a key is lost in a hospital, mechanical locks require reinstallation due to safety requirements. Digital locks do not; they can be easily reprogrammed. The locks developed by ILOQ use mechanical turning energy to power up the digital locks. Therefore the locking system does not need a power source and the system is very energy efficient. Although mechanical locks are less expensive at the time of procurement, digital locks are less expensive when you take into account the life-cycle of the product, i.e. the cost of reinstallation. With the digital locking system developed by iLOQ the public sector has saved substantial amounts of money and improved safety as well as energy efficiency. In addition, the company itself is now one of the fastest growing companies in Finland (revenue of EUR 10 million) and exports to Denmark, the Netherlands and Sweden.

For more information, see www.iloq.com/.

Source: Country response to OECD (2017), "OECD Survey on Strategic Procurement for innovation 2015", in *Public Procurement for Innovation: Good Practices and Strategies*, Annex C, OECD Publishing, Paris.

According to the results reported in the good practice cases, expectations were met for procurement for innovation (see Figure 2.3), especially considering the majority of the projects hadn't yet reached the projection stage at the time of reporting - which means that some reports need to be considered as emerging trends or preliminary results.

Most often, procurement for innovation helped to improve effectiveness and efficiency, as well as user satisfaction (16-17% in both cases). The improvements in both efficiency and effectiveness can be mostly attributed to energy savings, as found in the good practice cases presented below.

Figure 2.3. **Results achieved by implementing procurement for innovation good practice cases**

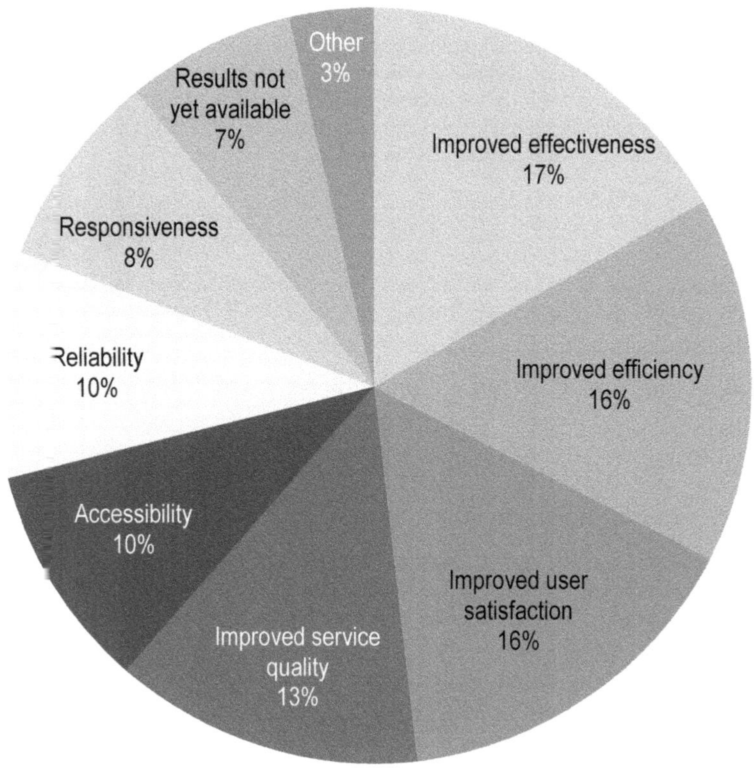

Note: Countries could provide multiple responses.

Source: Country responses to OECD (2017), "OECD Survey on Strategic Procurement for innovation 2015", in *Public Procurement for Innovation: Good Practices and Strategies*, Annex C, OECD Publishing, Paris.

In 13% of the good practice cases, procurement for innovation improved the quality of services, for example by improving the lighting provided by street lamps (as exemplified by France's and Poland's good practice submissions; see Boxes 2.5 and 2.6). Reliability and accessibility improved in about 10% of the cases. Services were made more accessible, for example by leveraging information technology (IT) systems. In the remainder of the cases, responsiveness (e.g. the ability to respond to different user needs or user views) increased, or results of the innovative procurement case were not yet available.

Box 2.5. Spotlight: Innovative practice in France

Liquid LEDs (bulbs for public lighting) (2014)

With the introduction of liquid light-emitting diodes (LEDs), the consumption of electricity is reduced up to 60%. Liquid LED lamps are more durable and have more functionality, such as video monitoring, possibility of changing the light intensity and colour/adding captors, alarms, global positioning system (GPS), among others. The bulbs are fully connected; most importantly, the candelabrum remains the same, so that there are no additional investments. The total cost of ownership (TCO) of liquid LED technology is 50% to 75% lower than that of normal LED technology.

For more information, see www.ugap.fr/actualites/innovation/actualites/soutien-de-lugap-aux-entreprises-innovantes-3-exemples_966189.html and www.ledliquidledex.com.

Source: Country response to OECD (2017), "OECD Survey on Strategic Procurement for innovation 2015", in *Public Procurement for Innovation: Good Practices and Strategies*, Annex C, OECD Publishing, Paris.

The second example on improving lighting is provided by Poland (see Box 2.6).

Box 2.6. Spotlight: Innovative practice in Poland

Hybrid lighting in the *Jarosław* commune (2010)

Within the framework of the hybrid lighting project 313 hybrid lamps that absorb energy from renewable sources (sunlight and wind) have been constructed on the territory of the *Jarosław* commune. Hybrid lamps were built along the communal roads in places lacking a traditional electricity supply network.

A single hybrid lamp is an independently functioning light source using renewables instead of receiving electricity from the grid. Its major advantage over conventional lighting is that it can be installed anywhere as it does not require connection to an electricity grid. Hence, there is no need for cable trenches and placement.

The hybrid lamp is composed of a lighting pole where a wind turbine (power of 600 W) is installed together with two photovoltaic cells (120 W each), two batteries (230 Ah each), a luminaire with a light source of 36 W, a solar controller and a wind controller. On average, the annual quantity of electricity produced with the use of wind energy is about 40 MWh, while the amount of electricity generated with the use of solar energy is about 62 MWh.

Implementing this innovative project achieved the following:

- improved general safety of residents: increase in safety of drivers and pedestrians on the roads and roadsides, reduced number of road accidents and collisions, fewer thefts and robberies
- addressed increased demand for electricity coming from the development of the economic and communal/municipal sectors
- addressed limited possibility of connecting to the grid system
- addressed efforts undertaken to limit pollution from energy production based on conventional sources linked to carbon dioxide and other dangerous emissions.

The official name of the project is "Modernisation of the street lighting in the *Jarosław* commune - Phase I: Lighting powered by renewable energy".

For more information, see http://jaroslaw.itl.pl/bip/index2.php?page=position2.php&id=2637&grp=3 and www.eko-gminy.pl/dobre-praktyki-archiwum/czerwiec-2011/222-jaroslaw-hybrydowe-oswietlenie

Source: Country response to OECD (2017), "OECD Survey on Strategic Procurement for innovation 2015", in *Public Procurement for Innovation: Good Practices and Strategies*, Annex C, OECD Publishing, Paris.

In over one-third (36%) of the good practice cases presented in this report (see Figure 2.4), the innovation related to the deliverable, i.e. the product or service was new or catered to a group that had not used it before. In less than one-third of the good practices cases (30%), the innovation was related to design or delivery, i.e. new approaches were used to design or deliver a product that had existed before. In one-fifth of the good practices (21%), the innovation related to communicating with users. In just over one-tenth (11%) of the cases, the innovation reorganised the work in the institution that implemented the programme. Under the category "Other", one country reported the creation of an entirely new knowledge base.

Figure 2.4. **Type of change resulting from an innovation cited in the good practice cases**

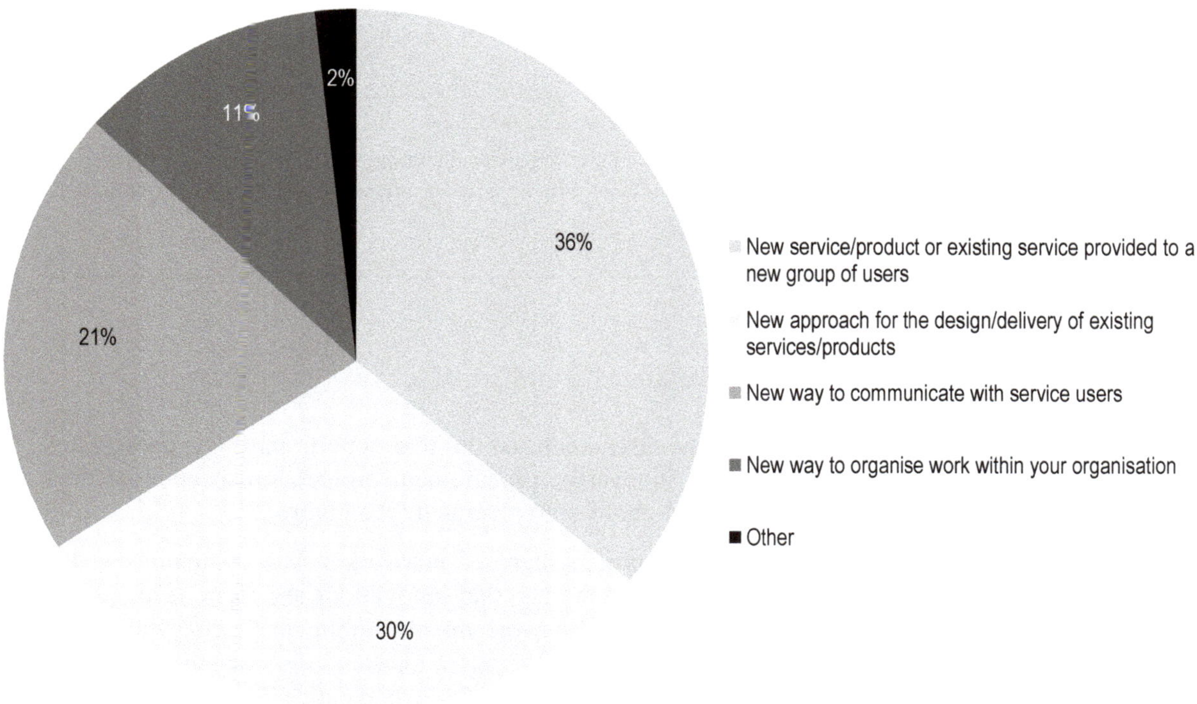

Note: Countries could provide multiple responses.

Source: Country responses to OECD (2017), "OECD Survey on Strategic Procurement for innovation 2015", in *Public Procurement for Innovation: Good Practices and Strategies*, Annex C, OECD Publishing, Paris.

Box 2.7 provides an example from Austria that illustrates how new approaches were implemented; in this case, a database to facilitate food purchases.

Box 2.7. Spotlight: Innovative practice in Austria (1)

Full range, (socially) sustainable food package with dynamic allergen indication (2014)

The food database and ordering system currently allows over 400 public organisations to order food with customised quality (organic, seasonal, agricultural products, different feedings [fetter methods], free of genetic engineering, fair-trade). This includes an optimised SME policy (focus on a regional structure of suppliers) as well as a full indication of all allergens. All suppliers have to update information on allergens and ingredients weekly. The information provided is then not only linked to the supplier, but also to the product, making it a smart database. In this database, public procurers can then choose which criteria to apply to their food (e.g. seasonal and gluten-free) and obtain a list of food that complies with the specifications. The database/framework agreement comprises all kinds of food (meat, dairy, dry food, soups) and is updated every four months.

For more information, see www.ioeb.at/ueber-ioeb-und-die-servicestelle/news/news-detail/public-procurement-of-innovation-award-2015-goes-to/.

Source: Country response to OECD (2017), "OECD Survey on Strategic Procurement for innovation 2015", in *Public Procurement for Innovation: Good Practices and Strategies*, Annex C, OECD Publishing, Paris.

Partners and beneficiaries of procurement for innovation

In most of the good practice cases featured in this report, partners contributed substantially to the success of the innovative procurement project, irrespective of their affiliation (e.g. buyers groups) or stakeholder involvement (see Figure 2.5).

The largest group of partners in procurement of innovation was companies in the private sector (33%); this seems related to the fact that innovations are often developed in this sector. The second largest group (27%) were mostly public institutions/bodies or other government entities. These include partnering ministries, government entities on other levels of government (e.g. municipalities) or state-owned enterprises or providers of government services (e.g. hospitals). The next largest group was academics and research bodies; almost a quarter (24%) of the partners in the good practice cases fell into this category. Civil society was relatively little involved; with only 8% of the partners in this category. Civil society can provide a valuable monitoring role in public procurement generally. Other partners included agencies, international organisations and IT providers (8%).

A good example of the benefits of partnerships in joint cross-border procurement is the EU-project Smart@Fire, funded by the European Commission, in which several European countries participated (see Box 2.8). This good practice case was submitted by Belgium, region of Flanders, and Hungary for the *Észak-Alföld* region (see also Box 2.2).

Figure 2.5. **Partners in procurement for innovation, according to good practice cases**

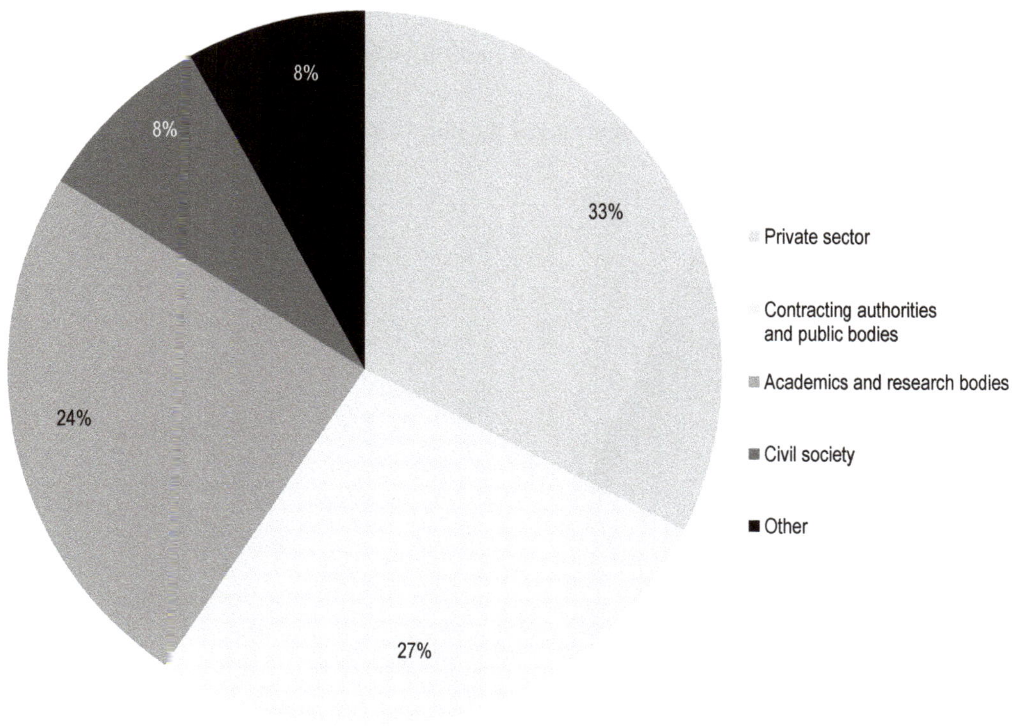

Note: Countries could provide multiple responses.

Source: Country responses to OECD (2017), "OECD Survey on Strategic Procurement for innovation 2015", *Public Procurement for Innovation: Good Practices and Strategies*, Annex C, OECD Publishing, Paris.

Box 2.8. **Example on partnerships: Participation in a European consortium**

The EU-PCP-Project Smart@Fire is related to the European Seventh Research Funding programme (FP7) and conducted as part of a larger consortium between different organisations, co-ordinated by the former Flemish Governmental Innovation Agency (IWT) in Belgium.

Each partner contributed a specific element of expertise. INNOVA, the regional development and innovation agency in Hungary's *Észak-Alföld* region, contributes experience in pre-commercial procurement (PCP) and event organisation. In addition, INNOVA takes on elements related to Hungary specifically, such as information dissemination and the representation of Hungarian firefighters and first responders. Other project partners (e.g. university, business associations and government institutions at the municipal and federal level) served as experts in technology or provided input on end-user requirements.

Note: For the European PCP example funded in FP7, a consortium of at least three partners from different countries was obligatory to participate in the chosen funding instrument.

Source: Country responses to OECD (2017), "OECD Survey on Strategic Procurement for innovation 2015", in *Public Procurement for Innovation: Good Practices and Strategies*, Annex C, OECD Publishing, Paris.

Figure 2.6 sets out the main beneficiaries of the good practice cases featured in this report. The two main beneficiary groups are citizens (28%) and public services (28%), followed by special groups, such as firefighters, patients and insurance companies (16%), suppliers (12%), SMEs (9%) and the general administration (7%).

Figure 2.6. **Main beneficiaries in procurement for innovation, according to good practice cases**

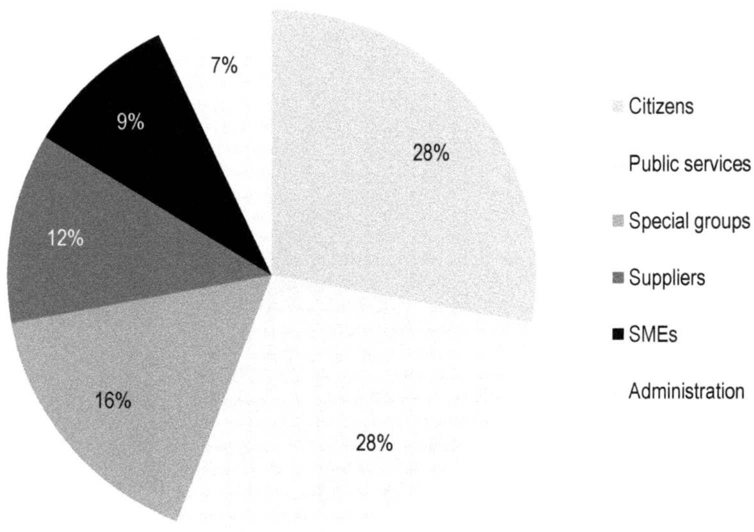

Note: Analysis of free-text responses.

Source: Country responses to OECD (2017), "OECD Survey on Strategic Procurement for innovation 2015", in *Public Procurement for Innovation: Good Practices and Strategies*, Annex C, OECD Publishing, Paris.

Main challenges implementing procurement for innovation

Countries have made progress in implementing procurement for innovation in recent years. However, as in every new policy area, a large part of the progress consists of overcoming stumbling blocks. Analysing the goals and comparing these starting points or intentions with the actual outcomes provides valuable insight. The obstacles are very often similar across countries; the challenges are also often interlinked, which makes it particularly difficult to address them. This section provides an overview of common obstacles countries need to address when developing or implementing policies for procurement for innovation. Figure 2.7 lists the categories of challenges most frequently encountered by countries in their attempts to implement procurement for innovation.

Figure 2.7. **Categories of main challenges to procurement for innovation**

Note: Analysis of free-text responses (categorised); n= numbers of responses provided.

Source: Country responses to OECD (2017), "OECD Survey on Strategic Procurement for innovation 2015", in *Public Procurement for Innovation: Good Practices and Strategies*, Annex C, OECD Publishing, Paris.

Some of the most frequently mentioned challenges are not related to issues usually associated with innovation or procurement for innovation, such as unknown technical complexities or higher costs for specialty orders. Rather, these challenges relate to aspects that are important in any administration or organisation. Management or co-ordination, for example, is cited more frequently than those areas that might – at first glance – pertain much more to procurement for innovation specifically, such as expertise of the unknown, innovative areas or financial resources to finance costly trials. Management and co-ordination would probably be identified as an everyday-challenge in any administration. Other key areas, such as monitoring, have not often been cited as challenging areas; at the same time, countries have ample room to increase their monitoring efforts. This indicates that countries are less aware than they should be of the benefits of monitoring in procurement for innovation. Measurement and monitoring can be crucial to implementing procurement for innovation as doing so highlights an innovation's strength and weaknesses. The achieved results are indicators for economic growth.

The challenges are linked: the organisational culture frequently depends on the "tone from the top" (management and leadership). The frequent quest for urgent fixes contradicts long-term strategies. Lack of funds also means fewer funds to spend on risk-mitigating measures, and fragmentation can result in a shortage of funds for a particular part of the system. Organisational culture determines the risk appetite, which is understood as an organisation's willingness to accept a certain level of risk. Lack of communication also means less knowledge exchange. Although a fragmented environment does not automatically result in bad co-ordination, fragmentation creates additional obstacles for co-ordination. The categories presented above were chosen to illustrate the issues at stake and assist countries in designing their procurement for innovation strategies to address some of these challenges (see Chapter 3).

The main challenges cited in the OECD Survey will be discussed in the following section. They are, in order of most- to least-reported by responding countries: 1) risk aversion; 2) management and co-ordination; 3) capacity (in terms of numbers and skills); 4) political support; 5) resistance to change; 6) legal and regulatory framework; 7) financial support; 8) fragmentation; 9) awareness-raising; 10) measurement systems and IT tools; and 11) time-related challenges.

Risk aversion

Challenges related to risk are one of the most common obstacles to implementing effective procurement for innovation. Bodies in charge of procurement associate procurement for innovation with higher risk, or they are generally averse to assume risks. Higher risk as such is not necessarily a problem: risk analysis can identify measures to mitigate risk, depending on an organisation's risk attitude. The fact that risk aversion is perceived as such an important obstacle to strategic procurement for innovation might be related to issues of capacity and organisational culture. Often, countries lack capacity, including expertise, to conduct risk assessments or pay for mitigating measures. Often, the organisational culture has a bias towards traditional measures: some countries acknowledged in their survey responses that risks might only be perceived as higher, when in fact procurement for innovation by itself is not always riskier than conventional procurement. Here again, the lack of resources and skills to manage risk successfully is a factor, i.e. to properly assess and mitigate risks related to procurement for innovation.

Management and co-ordination

Lack of management and co-ordination was also seen as a serious challenge. Weaknesses relate to strategic innovation management, knowledge of how to balance risks and benefits in public procurement needs assessments, early market consultation and/ or dialogue with stakeholders. In addition to the lack of management, the lack of communication or co-operation is also important to address. All of these issues are linked to fragmentation. Aside from a lack of co-ordination between different parts of a potentially fragmented system, lacking exchanges with external parties, such as industry associations or interest groups, constituted an obstacle to procurement for innovation as well. The three "Cs" of co-ordination, co-operation and communication were frequently mentioned together, providing an indication of how important it is to examine these three concepts together, taking into account the overall governance model.

Innovation often originates from fruitful collaboration rather than from isolation. In most countries, innovative ideas emerged from a dialogue between government entities and business, as well as end users/beneficiaries of the service. Countries such as Belgium, Canada and Norway (see Box 2.9), for example, undertook a consultative process that involved contracting government entities, businesses, and sometimes also experts and users. In Colombia, government agencies worked with consultancies to identify potential for innovative solutions. This highlights that innovation develops best when there is a collaborative dialogue between bodies with different perspectives. In fact, where this dialogue was present, the outcome was usually good.

Box 2.9. **Spotlight: Innovative practice in Norway**

Omsorg+ Kampen project within the National Program for Supplier Development (2008)

The procurement illustrates how extensive dialogue with the market can create space for new and innovative solutions. The purpose of the dialogue activities was to present the procurement plans, and to receive information on what kind of solutions were available on the market. This was followed up by one-to-one meetings with different potential suppliers. The case received considerable media attention, which mainly focused on the interaction between technology and human beings - especially elderly people with limited technology skills. Many Western European countries are facing the same challenge: a shrinking workforce will have to support a growing population of older people. Omsorg+ Kampen illustrates how welfare technology can meet this development by both saving costs and improving services.

For more information, see www.anskaffelser.no/sites/anskaffelser/files/gevinstanalyser_innovative_anskaffelser_16122014; www.bymisjon.no/Virksomheter/Kampen-omsorg/Hva-er-Omsorg/ and www.innovation-procurement.org/award/city-of-oslo.

Source: Country response to OECD (2017), "OECD Survey on Strategic Procurement for innovation 2015", in *Public Procurement for Innovation: Good Practices and Strategies*, Annex C, OECD Publishing, Paris.

Capacity (in terms of numbers and skills)

The lack of capacity in terms of numbers and/or skills is the third most frequently cited category of challenges. Challenges related to capacity pertain to two interlinked aspects: qualitative and quantitative needs of the public procurement workforce to conduct procurement for innovation. Qualitative aspects relate to issues like specialised skills and knowledge necessary to conduct public procurement for innovation. Quantitative aspects concern the sufficient availability of skilled staff to handle all procurement for innovation cases.

The issue of capacity is distinct from issues like cultural challenges: even if all officers are highly motivated to conduct procurement for innovation, staff might not know *how* to do so.

The lack of knowledge or skills can hinder procurement for innovation because it requires a strategic approach that goes beyond simple decision making based on simple criteria, such as price. Often, it is difficult to find people with these skills, e.g. people who are able to identify unmet needs, to balance innovation goals with primary procurement objectives, or who are able to build test environments for innovative prototypes or solutions. Some countries (e.g. Finland, Germany, Greece and Sweden) mentioned the lack of sufficient staff in general. Belgium highlighted the need for skilled staff for bottom-up implementation.

In addition, staff with specific skills is needed to manage and lead multi-disciplinary teams, consisting of professionals with a background in information technology, acquisition, financial management or law. Hiring highly-skilled staff is not only a challenge for the public agencies conducting procurement for innovation. Smaller companies as potential suppliers face similar challenges, because they often lack staff with knowledge of public procurement processes.

Aside from lack of skills, procurement for innovation can also be hindered due to lack of officials available or able to conduct innovative projects, even where the few officials available are highly skilled. The two dimensions are closely related, and also have a financial aspect: hiring staff might be costly and hiring skilled staff can be even more so. The availability of highly knowledgeable staff to support procurement for innovation is often a function of the availability of financial resources. Hiring highly qualified staff is costly (see the challenge of financial support, below).

Political support

The existence of a national procurement for innovation action plan can be regarded as a baseline condition for applying new procurement schemes. What is needed is smart policy development, strong political will and commitment as well as experimental innovation policy approaches and a clear and consistent policy framework. Policy measures can take this into account by setting up initiatives to increase knowledge on procurement for innovation practice, provide guidance and offer financing opportunities.

Resistance to change

Another major hurdle to procurement for innovation is related to the organisational culture in the procurement function on all entity levels. Public procurement officials may resist change or have attitudes that counter creative, forward-thinking attitudes required for trying new and innovative approaches. Motivation to apply innovative approaches is missing; staff lack interest in innovative solutions. This aspect comprises elements such as political and administrative leadership or financial support for procurement for innovation, motivation of the staff, and awareness that innovative solutions can and should be sought.

This aspect is critical and difficult to manage: an organisational culture is much harder to control; simple, centralised approaches such as increased financing will not work. At the same time, the organisational culture appears to be one of the most important success factors supporting innovation. Often, realising opportunities for procurement for innovation depends on the individual who either thinks creatively to find a solution, or feels confident to take up an unusual offer. How to handle frustration was also mentioned as an issue – new processes present greater opportunities for failure. Often, individual motivation is linked to the "tone from the top": only where leadership backs creative solutions and encourages individuals to seek innovations does procurement for innovation materialise.

Legal and regulatory framework

Legal provisions often provide obstacles to procurement for innovation: either because the traditional public procurement framework in place hampers innovative solutions, or because specific regulations guiding procurement for innovation are missing. Albeit not restrictive, some existing frameworks were interpreted as restrictive – and in opposition to - procurement for innovation.

This last aspect on the interpretation of the law supports the above-mentioned challenges related to risk aversion, especially when the legal framework provides limited clarity. In these cases, public officials might fear litigation and will refrain from actions that could be interpreted ambiguously in court.

Financial support

Innovation is often the outcome of a cost-intensive process: there are higher risks involved and it is central to experiment, which consumes further resources. Sufficient resources are important for all organisations involved in undertaking the procurement for innovation process (e.g. an agency, but also on the side of the company that is supplying the innovative product or service). Frequently, countries do not have sufficient financial support available to conduct procurement for innovation. Procurement for innovation is often associated with higher-than-usual costs (whether perceived or real). Aside from a sufficient level of funds, countries expressed the need for dedicated funds to be used specifically for procurement for innovation. In the European Union, these needs have been addressed for example by the PCP and PPI funding in the European Union's Research and Innovation Programme Horizon 2020 offered by the European Commission. In the success stories featured in this report, sufficient financial resources were one of the key facilitating factors.

Fragmentation

Procurement for innovation activities are often performed by different institutions that play different roles: responsibilities are allocated only to part(s) of the activity. For example, responsibilities of one institution are related only to procurement elements, while other institutions hold the responsibility related to the "content", i.e. the good or service that is being procured. Strategic procurement for innovation is therefore often hindered by fragmentation (or low integration) in different dimensions. This includes the fragmentation of governance systems and administrations, for example due to different governance levels, federal structures or several institutions taking part in procurement for innovation. Accordingly, unclear responsibilities and fragmented decision-making power were named. Also mentioned was fragmentation of funds and of funding sources for procurement for innovation.

Awareness-raising

Despite the recent increase in attention to the strategic function of procurement for innovation, there is still a lack of awareness around procurement for innovation. Procurers, as well as potential suppliers, are unaware of the role that public procurement has with regards to procurement for innovation, and do not know that public procurement might constitute demand for innovative products. In addition, there might be a lack of knowledge about the processes related to procurement for innovation. Some countries started awareness-raising initiatives (e.g. Austria, Belgium, France, Germany, Malta and the United Kingdom). This ensured awareness of their innovation programmes and potential benefits across departments and public bodies, and was complemented by training. The latter emphasises the link between awareness and the capacity (in terms of numbers and skills) challenge.

Measurement systems and IT tools

Measuring impact of public procurement for innovation activities appears to be an area that countries pay little attention to. Only 44% (15 countries, a little less than half of all responding countries) have a system in place to measure the impact of their procurement for innovation policies. This finding provides motivation for recent OECD work on measuring the link between public procurement and innovation (Appelt and Galindo-Rueda, 2016).

Countries place too little focus on the issue of monitoring, despite the demonstrated benefits and results. The low level of monitoring does not seem to be considered harmful to procurement for innovation efforts. Yet, being able to demonstrate the benefits of an innovative approach (as opposed to a traditional approach) can support successful implementation of procurement for innovation. Only one country considered lack of capacity to monitor the results or the impact of procurement for innovation as an obstacle to successful procurement for innovation. Four countries expressed intent to improve their systems for measurement when asked for solutions. Estonia, for example, is working on implementing a system for measuring innovative procurements in an e-procurement database.

Where impact evaluations of procurement for innovation are conducted, most are done in the form of evaluation studies (34%) or state-of-play studies (22%). Some 11% are conducted as impact assessments. A third of the evaluations (33%) are conducted in other forms, including as part of academic research, surveys of government supplies, feasibility studies or specialised one-off publications (see Figure 2.8).

Figure 2.8. **Instruments used to measure the impact of procurement for innovation activities**

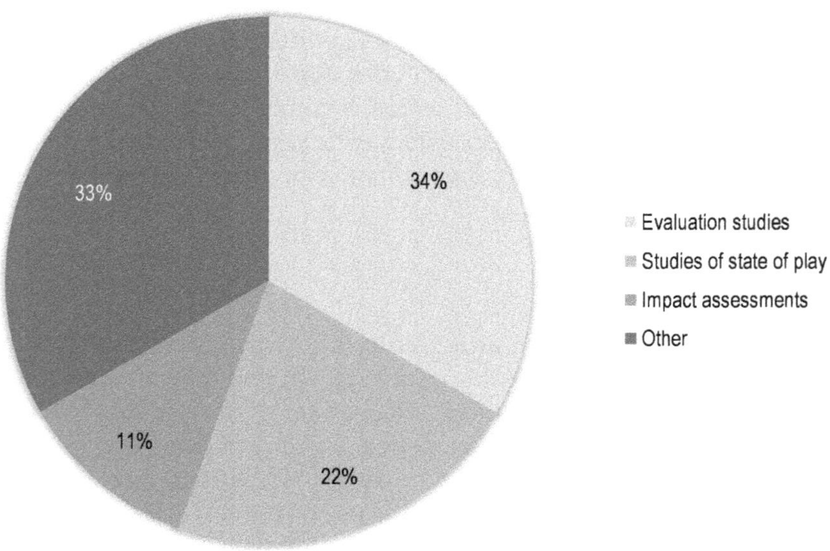

Note: Calculated on the basis of 44% "yes" responses to the question on having a system in place to measure the impact of actions related to procurement for innovation. Countries could provide multiple responses.

Source: Country responses to OECD (2017), "OECD Survey on Strategic Procurement for innovation 2015", in *Public Procurement for Innovation: Good Practices and Strategies*, Annex C, OECD Publishing, Paris.

Those countries that have already implemented monitoring systems employ a wide range of tools with the aim of measuring the impact of their procurement for innovation policies. Such tools include:

- surveys
- external or independent reviews (e.g. conducted by auditing firms)
- combined interim and *ex post* evaluations
- leveraging existing tools, such as statistics or e-procurement systems by adding a marker highlighting innovative procurement cases
- one-off project-related evaluations or one-off evaluations on the innovation strategy in general
- general reviews, evaluations and monitoring exercises of public procurement practices, including elements related to innovation.

As mentioned, the majority of responding countries (56%) have no system in place to measure the impact of procurement for innovation. Half of the countries that do not monitor impact provided reasons for not conducting monitoring: overwhelmingly, it relates to the novelty of procurement for innovation. On the one hand, countries stated that there was no policy to monitor impact – for example because procurement for innovation was conducted in one-off projects. On the other hand, when the countries had a dedicated procurement for innovation policy, countries stated that the policy was too new to be monitored in a meaningful way, or that they planned on doing so in the future. One country responded that monitoring at the national level was not possible because relevant data was only gathered at the regional level. Another country is planning to conduct monitoring of its procurement for innovation strategy as part of the monitoring related to its National Reform Programme under the European Union's Europe 2020 Strategy.[3]

Targets are common means to support procurement for innovation. Less than one-third of all responding countries (11 countries) have set targets related to procurement for innovation – mostly prescribing a percentage of public procurement value that should be conducted under procurement for innovation aspects. All countries that have set a target also follow up to measure it. Only three countries have reached their target, i.e. only around one-quarter of those countries setting a target have reached it. Less than one-third of the responding countries have not set a target.

The examples presented in Box 2.10 show that countries sometimes set a target to be spent in general on innovation, or explicitly on procurement for innovation.

Box 2.10. Examples of procurement for innovation targets

Examples of countries with quantitative procurement for innovation targets:

- Government programme 2015 includes a 5% target for innovative public procurement (Finland).
- SMEs must reach 2% of procurement for innovation by 2020 (France).
- An ambition of 2.5% to be spent on innovation (Netherlands).
- A target of 3% in new investment for procurement for innovation (Spain).

Box 2.10. **Examples of procurement for innovation targets** *(continued)*

- Central/local governments and public enterprises should fulfil 20% of their procurement of the specific product type for which new-technology, certified products are available (Korea).

Examples of countries with qualitative procurement for innovation targets:

- indicative targets to stimulate procurement for innovation (Netherlands, Belgium/Flanders)
- promotion of effective and innovative public procurement (Denmark)
- increasing share of domestic firms in high-tech-sectors in procurement for innovation (Turkey).

Source: Country responses to OECD (2017), "OECD Survey on Strategic Procurement for innovation 2015", in *Public Procurement for Innovation: Good Practices and Strategies*, Annex C, OECD Publishing, Paris.

Time-related challenges

Countries also emphasised challenges related to insufficient time available. This category of challenges comprises two aspects: 1) insufficient time available to procurement officials to conduct an innovative public procurement process (i.e. time for planning issues, such as needs assessment, market consultation, patent research, corrections, experiments, etc.); and 2) favouring fast results over solutions that take time.

Others

Challenges to procurement for innovation reported less frequently included:

- issues related to intellectual property rights (IPR)
- a recurring pattern that pilots (one-off cases) do not become general practice, a standard process or day-to-day business despite efforts
- opposing interests of parties involved in the procurement for innovation process.

The protection of intellectual property rights was mentioned by one country, although IPR protection plays a big role in the understanding of risk and benefit sharing and as part of competition rules. For instance, the specific requirements for the R&D services exemption of the EU Public Procurement Directives for Pre-Commercial Procurement (PCP) have to be respected, so that the sharing of IPR rights in PCP takes place according to market conditions (European Commission, 2007).

Challenges and key lessons learned - measures

This section analyses how countries responded to the challenges encountered in connection with their procurement for innovation initiatives. Countries reported to have met several challenges and to have acquired a wealth of lessons learned – touching on most aspects of the public procurement process. Interestingly, there is a discrepancy between what kind of actions the most-frequently identified challenges would require, and the actual measures chosen by countries.

Most frequently, procurement for innovation was stalled by

- risk aversion,
- management and leadership problems,
- lack of professionalisation (capacity and skills), followed by
- political support and cultural challenges in the public procurement workforce (see Figure 2.7).

As demonstrated by countries, possible responses include first, providing more training and education for procurement officials, and second, improving the legal framework. The latter was mentioned less frequently. Countries undertook measures to improve the innovation culture; however, these activities were rarely implemented and not as frequently as would be adequate to counter the challenges identified.

One caveat: The question did not specifically ask for measures, but rather asked whether obstacles were overcome. Therefore, it is likely that countries did not report measures even if the countries had already implemented them.

Box 2.11 presents a list of the most frequently used measures to overcome procurement for innovation challenges.

Box 2.11. Most frequently used measures to overcome challenges

- Professionalisation: Training, education
- Legal framework: Changing laws, introducing regulations
- Culture: Increasing internal awareness about procurement for innovation
- Funding: Increasing or solidifying financial resources for procurement for innovation
- Outreach: Specific measures to engage stakeholders
- Supplier awareness: Support/education for potential suppliers
- Monitoring: Introduction of monitoring or evaluation requirements/exercises

Note: EU countries were obliged to transpose the 2014 EU Procurement Directive into national law.

Source: Country responses to OECD (2017), "OECD Survey on Strategic Procurement for innovation 2015", in *Public Procurement for Innovation: Good Practices and Strategies*, Annex C, OECD Publishing, Paris.

Other noteworthy, successful methods include the introduction of a target (Finland, 5% of total procurement) and the organisation of workshops to improve co-ordination between different parts of the public procurement system (Turkey). Among other measures, New Zealand introduced "Government Rules of Sourcing", which provide a flexible and supportive environment for good procurement practices, guides, tools and templates. Some accompanying actions are capability development; improving professional standards in government ministries and departments; reviews of ministries' and departments' procurement practices; and development of procurement capability. Box 2.12 provides an example of a comprehensive set of measures implemented to overcome obstacles related to procurement for innovation in the United States.

Box 2.12. United States' operationalisation of lessons learned in procurement for innovation

The United States has listed more than a dozen concrete measures to address challenges to procurement for innovation. The list is comprehensive and covers many aspects of the procurement cycle:

- restructuring and connecting data storage
- usage of cloud services
- more attractive career paths
- redefining of roles and new requirements for staffing teams
- knowledge sharing across government and outside
- co-operation with universities
- guidance and templates
- changes to rules on budget and spending
- "myth-busters" campaign to address the reluctance for exchange between the private and public sectors.

Note: See the United States' country factsheet example in this report (Annex A).

Source: Country response to OECD (2017), "OECD Survey on Strategic Procurement for innovation 2015", in *Public Procurement for Innovation: Good Practices and Strategies*, Annex C, OECD Publishing, Paris.

In addition, the following measures have proven effective to promote procurement for innovation. These measures are also in line with the suggested OECD Framework to Promote the Strategic Use of Public Procurement for Innovation, as proposed in Chapter 3. Most of the measures respond to at least one challenge reported, but they typically work better together and address several challenges to procurement for innovation at once.

Communication and co-ordination

The areas that were most frequently underestimated related to communication, outreach and awareness – both internally within the public procurement function and externally with partners. Extensive communication and co-ordination proves important for the success of innovation policy and practice. This aspect is closely linked to the challenges described in relation to management and leadership in administration and to a culture of innovation. Procurement for innovation thrives where:

- Good communication and dissemination about the different aspects of innovation are in place.

- Co-ordination between different levels of government (horizontal and vertical) exists.
- General benefits or project specifications, etc. are present.
- Procurement officers maintain a market consultation and a dialogue with potential suppliers.
- Risk and benefit balance is known and the acceptance of innovation public procurement is high.
- Partnerships with relevant stakeholders are formed.
- Existing business associations or stakeholder groups are engaged in support of procurement for innovation.
- Co-ordination between different branches of the public procurement process is good.

In turn, where these aspects were not present, procurement for innovation efforts proved to be difficult.

Strategy

Leadership and strategy proved to be key factors for successful implementation of procurement for innovation. In many of the cases featured in this report, it was helpful if the strategy was clearly embedded in a dedicated procurement for innovation strategy, and was incorporated in any secondary policy objectives that the country sets for itself. Strategies at higher levels helped to co-ordinate between different lower levels of government when it came to procurement for innovation, because a common direction eliminated disagreements. Overarching strategies appear to be a common, well-placed way to show political support for procurement for innovation, the lack of which having been identified as one of the main challenges.

Change management

For many countries, procurement for innovation is a challenge mostly because it is something new and requires adaptation. Change management can respond to this issue, and can help countries implement other measures taken to address challenges. Experience shows that countries underestimated the length of time needed for agencies to conduct innovative procurement procedures. It also took unexpectedly longer to arrive at a culture that accepted the "new ways". Two countries successfully responded by introducing measures to highlight the benefits of procurement for innovation both to their public institutions and the private sector. This facilitated the acceptance of procurement for innovation.

Developing technical solutions

An aspect that was often overlooked in the design of policies, but has practical relevance, pertained to technical challenges. For example, procurement processes had sub-optimal outcomes because bidders entered the process with an insufficient level of technical maturity, which means that the product offered was not advanced enough and therefore did not meet the requirements. In response, functional specifications can be introduced to ensure that the offered products meet the needs. At the same time, specific requirements often rendered offers less innovative as there was little room for creative solutions. For example, Finland mentioned that it was ideal to procure a solution, not a standard: "One should purchase **results**, not **standards**,…" saying, i.e. not procurement of 'mechanical locks' but one should procure the 'best solution for locking our school."

Reducing the burden

Heavy procurement processes have been known to discourage companies from bidding. In the same vein, burdensome testing requirements increased the workload on the part of the procuring entity. Flat fees are one possible way to incentivise innovative solutions; these fees also simplify the procurement process. An innovative procurement initiative was most successful where the involved parties were able to realise a win-win scenario, with a satisfying solution for the public and a profit for the company.

Finally, countries reported a need for good documentation – including the mistakes in the procurement process – so as to learn and improve in the next process.

Combining secondary policy objectives and coherence

Previously, public procurement had a one-dimensional mandate to procure goods and services for the public sector, following the criteria of "best value for money". Depending on the strategic priority area, secondary policy objectives are increasingly supporting other governmental policy goals, i.e. sustainability, innovation, environmental standards, support for small and medium-sized enterprises and socio-economic priorities. Considering combined secondary policy objectives, the public sector could achieve coherence and added value in the form of avoiding unnecessary duplication, and using synergy effects for inclusive growth.

To reach these additional targets, public procurement can be used as a "lever" that offers high market volumes (for the public sector) and good controllability. The awareness and readiness to add additional objectives to the tender specifications are not new, especially in the field of sustainability. But combining secondary policy objectives with each other, to take better advantage of the above-mentioned power of public procurement as a strategic "lever", is not a simple task. Some countries have, however, embarked upon high-level co-operation and co-ordination, departing from silo thinking, as in the case of Sweden (see Box 2.13).

Box 2.13. Spotlight: Innovative practice in Sweden

Electrified roads (2013)

The purpose of this project is to create a knowledge base for industrial, academic and political decisions on possible future development and implementation of electrified roads in the Swedish road traffic system for heavy traffic. The demand for transport is increasing. Sweden does not have capacity in the railroad system to meet the demands. Railroads are expensive, and time is scarce to expand the railroad system. However, Sweden has capacity in the road system. Electrified light traffic is not critical and the implementation of such vehicles is in progress. However, there is no sufficiently good sustainable solution for heavy traffic.

For more information, see www.trafikverket.se/en/startpage/about-us/news/2016/2016-06/first-electric-road-in-sweden-inaugurated/.

Source: Country response to OECD (2017), "OECD Survey on Strategic Procurement for innovation 2015", in *Public Procurement for Innovation: Good Practices and Strategies*, Annex C, OECD Publishing, Paris.

Innovation and sustainable green growth

In most countries, public procurement has graduated from a more limited administrative procedure to "smart" procurement management, including strategic procurement for innovation. An effective government strategy for procurement for innovation and the right policy mix are the most important conditions for the implementation and high impact of an innovation-friendly environment.

New concepts are arising to define the combination of secondary policy objectives. Finland included innovation through public procurement in its national "Cleantech" policy and corresponding action plans, which aim to achieve clean technologies. It is regarded as part of Finland's general environmental protection efforts, as it means less pollution associated with every product that is developed and brought to market. This does not necessarily imply the development of new innovative solutions, but it does provide the market with existing environmentally friendly products that are also innovative (e.g. LEDs, solar energy infrastructure for households and mobility or low-carbon products to reduce CO_2 emissions).

For the EU Action Plan for "Circular Economy" (European Commission, 2016b), which describes a new perspective on how to use non-renewable resources for the production of goods, aiming to minimise the burden of contamination of nature by separating and exploiting reusable components after the end of the product's life-cycle, procurement for innovation can play an important role in form of "Circular Procurement".

Box 2.14 presents a good practice example from Colombia on combining innovation, socio-economic and environmental secondary policy objectives.

Box 2.14. Spotlight: Innovative practice in Colombia

The identification of cross-cutting secondary policy objectives was co-ordinated by Colombia Compra Eficiente (the National Public Procurement Agency) with the participation of other government agencies in need of innovative products and services. Several projects were developed, including the following pilots and ongoing initiative of the city of Medellin through *Ruta N*:

1. The Ministry of Information and Communication Technology (MinTic) started the contractual process for a laboratory to develop management IT skills and information security. The purpose is to recreate scenarios to run security tests, cybersecurity and develop research in IT and information privacy. In this case the direct beneficiary is the ministry itself. Indirectly, other government agencies benefit from the developments made in the test laboratory. It is important to run tests of fictional scenarios to verify the quality of the current tools to protect information and to test new developments. Through this process, MinTic tests different security aspects in safe and controlled circumstances. In addition, Colombia seeks to encourage its youth to study IT; procurement for innovation in the IT sector is a good incentive for doing so.
2. The National Agency for Overcoming Extreme Poverty (ANSPE) launched a programme that allows extremely poor communities to generate income by developing soft skills. Innovation is created from the need to combine different strategies (educational, technological, etc.) to identify opportunities and knowledge to overcome poverty. The direct beneficiaries are the communities experiencing extreme poverty; in an indirect way, the general population will experience improvements in their socio-economic situations. The Millennium Development Goals established specific goals towards overcoming poverty and moving society towards equality. ANSPE identified the need to develop a programme that mixes technology and knowledge as a way to provide populations in extreme poverty with the tools and knowledge to accomplish this.
3. The Empresas Públicas de Medellín (EPM - State owned enterprises of Medellín) undertook a project to reduce wastewater by identifying locations of leaks in real time. The project looks for a system to control costs to be charged to the users within monthly bills and to create awareness mechanisms related to water consumption impact. This product has not been built yet in Colombia. The direct beneficiaries are EPM and consumers. EPM will reduce its costs of finding water leaks; consumers will not pay for water lost. In fact, the costs for water lost due to leaks are transferred to consumers and has an important environmental impact. This represents awareness of water consumption and cost efficiencies, and is hard to achieve through regular procedures.

For more information, see www.colombiacompra.gov.co/compra-publica-innovadora/introduccion

Source: Country response to OECD (2017), "OECD Survey on Strategic Procurement for innovation 2015", in *Public Procurement for Innovation: Good Practices and Strategies*, Annex C, OECD Publishing, Paris.

Some OECD Survey respondents placed a focus on eco-innovation towards sustainable development (e.g. Austria, Ireland and Portugal), which is in line with the Eco-innovation Action Plan (European Commission, 2011) of the European Union. Estonia and the Netherlands offer examples for driving energy-efficient innovation though procurement as an opportunity for combining secondary policy objectives, resulting from procurement for innovation. The Estonian Electromobility Program (ELMO) launched in 2011 by the government of the Republic of Estonia, makes electric vehicles available to rent to the public for a short period of time, using a new, country-wide charging network for electric cars, to promote the use of electric vehicles.[4] Furthermore, the Erasmus University Medical Centre in the Netherlands initiated the procurement Robotic Bed-washing Facility (see Box 2.15).

Box 2.15. **Spotlight: Innovative practice in the Netherlands**

EU project EcoQUIP - Bed-washing Facility (2015)

Erasmus University Medical Centre initiated the procurement "Robotic Bed-washing Facility". In this procurement, the Erasmus University Medical Centre asked the market to design a more cost-efficient solution to disinfect 70 000 hospital beds and mattresses annually in a way that would also use less energy and water.

Erasmus University Medical Centre used the Forward Commitment Procurement principles that included a series of market soundings that stimulated cross supply chain interaction, a competitive dialogue and outcome-based requirements.

In the selection phase less emphasis was put on past experiences with bed-washing facilities than in regular procurements. The contract was won by IMS Medical. This company offered a robotic solution that includes high-precision cleaning robots from the automotive industry. The costs per bed were lowered by 35% and the CO_2 footprint was lowered by 65%. Furthermore, patients of the Erasmus University Medical Centre have cleaner beds and cleaning quality has become more consistent.

The focus on the problem in this procurement instead of the suggested solution made it such that IMS Medical could think outside the box.

Large possibilities are available to export the wash unit to one of the other 15 000 European hospitals. Currently IMS reports many followers and has had a number of interested hospitals. When the facility is proven operational IMS Medical will start a larger international marketing campaign.

Furthermore, the system can be applied for the cleaning and disinfection of all equipment with standard dimensions that need cleaning in large volumes. The budget for the procurement was EUR 1 million. Erasmus University Medical Centre conducted this procurement with support and European funds via the LCB Healthcare and EcoQuip project. With the procurement IMS Medical was able to increase its staff by 25%. Furthermore, around five additional jobs were created at the machine builder.

IMS Medical is an SME from Grootebroek, a small town in the Dutch province of North Holland. IMS Medical is active in the medical sector. It has a turnover of around EUR 1 million and eight employees.

Erasmus University Medical Centre is based in Rotterdam, Netherlands. With 1 320 beds, it is one of the largest hospitals in the Netherlands.

For more information, see www.erasmusmc.nl/corp_home/corp_news-center/2015/2015-11/bedden.duurzaam.gereinigd.door.robots/.

Source: Country response to OECD (2017), "OECD Survey on Strategic Procurement for innovation 2015", in *Public Procurement for Innovation: Good Practices and Strategies*, Annex C, OECD Publishing, Paris.

Innovation and the development of SMEs

Another model for combining secondary policy objectives can be seen in the innovation capacity of SMEs, which are in fact the working environment for the majority of the world's population. Data show that 95% of enterprises in OECD countries are considered SMEs, accounting for 60% of private-sector employment worldwide (Ayyagari, Demirgüç-Kunt and Maksimovic, 2011). SMEs are often regarded as the nucleus of innovation. One reason may be that in small companies there is little organisational and hierarchical overhead, and new ideas can be tested spontaneously without passing long procedures of permission seeking. Another reason is that those with an innovative idea found a company to develop this idea as a business – which initially has only a limited scope.

The OECD Survey confirms the link between SME support and procurement for innovation. In almost every country some kind of SME support is already in place on the innovation or employment policy agenda; this might be regarded as an advantage when combining secondary policy objectives. The support for SMEs to participate in procurement for innovation can be a direct financial incentive, a guarantee or an indirect measure like a SME participation quote, administrative assistance, training offers or other access-facilitating measures for public tenders on different levels. A combination of different kinds of measures has been reported by countries; see the examples from Canada, Czech Republic, Ireland and Korea, as presented in Boxes 2.16 through 2.19.

Box 2.16. **Spotlight: Innovative practice in Canada**

Build in Canada Innovation Program (BCIP, 2010-12)
(Formerly known as the Canadian Innovation Commercialization Program)

With the collaborative efforts of multiple government organisations and industry partners, the BCIP helps innovators bridge the pre-commercialisation gap by helping them move their innovations from the lab to the marketplace through testing in operational environments across government. The BCIP awards contracts to entrepreneurs with pre-commercial innovations through an open, transparent, competitive and fair procurement process for their testing within the Canadian federal government. The programme facilitates testing opportunities within the federal government with testing departments being required to provide feedback to entrepreneurs on the performance of their goods or services. In doing so, the BCIP provides innovators with the opportunity to enter the marketplace with a successful application of their new goods and services. With the help of the Office of Small and Medium Enterprise's network of Public Works and Government Services Canada, the BCIP also provides information on how to do business with the government of Canada.

For more information, see https://buyandsell.gc.ca/initiatives-and-programs/build-in-canada-innovation-program-bcip.

Source: Country response to OECD (2017), "OECD Survey on Strategic Procurement for innovation 2015", in *Public Procurement for Innovation: Good Practices and Strategies*, Annex C, OECD Publishing, Paris.

In addition to the Canadian "Innovation Commercialization Program", the Czech Republic provides a good practice example of a project where a need was identified to find innovative ways to use up-to-date techniques for archive operations. The project also employs the procedure of using a framework agreement with more SMEs.

Box 2.17. **Spotlight: Innovative practice in the Czech Republic**

New product using new processes for the use of state archives (2014)

In line with the programme BETA, the project supports the research and development of administrative and organisational solutions for the archive management for state archives. There is a need to identify innovative approaches using up-to-date techniques for archives operations. The project employs the procedure of a framework agreement with more SMEs and proceeds in steps to the final 2 and more results/prototypes. The project is still running.

The project aims to develop entirely new products using new processes for the use of state archives. There will be specific software modules developed at the end of the project. The innovation lies in the product as well as in the processes used.

For more information, see www.tacr.cz/index.php/cz/metoda-pcp.html.

Source: Country response to OECD (2017), "OECD Survey on Strategic Procurement for innovation 2015", in *Public Procurement for Innovation: Good Practices and Strategies*, Annex C, OECD Publishing, Paris.

The following example from Ireland combines SME support with sustainability and energy-saving goals and uses a pre-commercial funding scheme.

Box 2.18. **Spotlight: Innovative practice in Ireland**

Small Business Innovation Research (SBIR) - Electric vehicle smart charging (2014)

Under the government's Action Plan for Jobs for 2014, the government committed to introducing, on a pilot basis, a SBIR programme to provide opportunities for innovative solutions to be developed to meet the needs of public bodies. In this regard on 30 June 2014, the Sustainable Energy Authority of Ireland (SEAI), in co-operation with Enterprise Ireland and Electricity Supply Board (ESB), launched Ireland's first Small Business Innovation Research (SBIR) competition. SBIR falls under the category of pre-commercial procurement (PCP).

For more information, see www.seai.ie/SBIR; www.seai.ie/SBIR/SBIR-EV-Smart-Charge-Frequently-Asked-Questions.pdf; www.seai.ie/SBIR/SBIR-EV-SMART-CHARGE-CHALLENGE-The-Competition-Process-Nathalie-Sheridan.pdf; and www.seai.ie/News_Events/Press_Releases/2015/SBIR-Phase-1-winners-EV-chargers-Rlse-Final-2Apr2015.pdf.

Source: Country response to OECD (2017), "OECD Survey on Strategic Procurement for innovation 2015", in *Public Procurement for Innovation: Good Practices and Strategies*, Annex C, OECD Publishing, Paris.

Korea started combing the policy goals of procurement for innovation and SME development in 2004 as a national programme for the support of innovative SMEs; it then combined these goals with the standardisation of the procurement process.

Box 2.19. Spotlight: Innovative practice in Korea

Excellent Government Supply Products Program (2005)

This is operated by the Public Procurement Service (PPS). Each year, PPS selects SME products with excellent technology and quality, and certifies them as "Excellent Government Supply Products". PPS establishes framework contracts for the certified products, and makes them available at the Online Shopping Mall within the government-wide e-procurement portal (Korea Online e-Procurement System). Public entities can directly place online orders for these products without going through the bidding process. A separate aisle for these Excellent Government Supply Products has been established within the Online Shopping Mall, to give the products high visibility, and PPS also promotes them through Korea Public Procurement Expo and printed catalogues. Once selected, the Excellent Product can maintain the status for three years. To ensure that the selection is objective and fair, PPS invites over 700 external experts to evaluate the products. PPS annually certifies around 300 products for three years, maintaining about 1 000 products as Excellent Government Supply Products. In 2014, the total public sector purchase of the Excellent Products was approximately KRW 2.1 trillion (about EUR 1.5 billion).

Source: Country response to OECD (2017), "OECD Survey on Strategic Procurement for innovation 2015", in *Public Procurement for Innovation: Good Practices and Strategies*, Annex C, OECD Publishing, Paris.

Conclusions

Encouragingly, the majority of OECD Survey responding countries (almost 80%) undertake specific activities related to procurement for innovation. However, half of the countries do not have a stand-alone action plan or an action plan as part of a specific strategy to support procurement for innovation. Moreover, less than half of the countries (44%) have a system in place to measure the impact of procurement for innovation and only one-third of them have specified a target for strategic procurement for innovation, which clearly indicates that there is still room for improvement. Sound measurement systems are crucial for the evaluation of the national procurement for innovation strategies and improvements of the return of investments and social benefits. Apart from these implications for data-and indicator-related work, considerations with regard to harmonisation in monitoring models and knowledge sharing are required.

The OECD Survey highlighted a number of challenges that impede procurement for innovation. Most importantly, these challenges include risk aversion, management and leadership issues at administration level, professionalisation, political support and the organisational culture in the bodies in charge of procurement for innovation. Further important challenges were seen in a lack of willingness to change, lack of personnel capacity, insufficient financial support and the lack of a legal framework, followed by fragmentation.

Factors determining the success of procurement for innovation projects hinge on an organisational culture and leadership in administration that embraces innovation. Countries need to break the cycle in which lack of capacity creates the impression of higher risks, and the (perceived) higher risk affects the organisation's capacity to engage in innovative approaches. This requires investments in skills and competencies in public administration, as well as organisational and cultural change.

Good communication and co-ordination is part of that mix to stimulate procurers. For instance, knowledge exchange and trainings can help to better calculate the balance between risks and benefits. Additionally, public procurement agencies require sufficient resources to implement procurement for innovation processes. Public procurement agencies partner with a range of external actors, most importantly the private sector.

Therefore, alignment needs to be achieved not only across levels of government, but also between the public sector and suppliers to implement procurement for innovation successfully. Exchange with potential suppliers (as early as possible when starting an innovation-procurement project), and involvement of end users and non-governmental stakeholders (to ensure an optimum of benefits for all), prove vital in creating innovations. Additional gains will result from the synergy between other secondary procurement goals like sustainability, green growth and the development of SMEs.

Notes

1. *Note by Turkey*: The information in this document with reference to "Cyprus" relates to the southern part of the Island. There is no single authority representing both Turkish and Greek Cypriot people on the Island. Turkey recognises the Turkish Republic of Northern Cyprus (TRNC). Until a lasting and equitable solution is found within the context of the United Nations, Turkey shall preserve its position concerning the "Cyprus issue".

   Note by all the European Union Member States of the OECD and the European Union: The Republic of Cyprus is recognised by all members of the United Nations with the exception of Turkey. The information in this document relates to the area under the effective control of the Government of the Republic of Cyprus.

2. Country spotlights are presented in alphabetical order by country when more than one is presented.

3. In 2010, the European Commission launched the European Union's Europe 2020 Strategy to create the conditions for smart, sustainable and inclusive growth. Under this strategy EU member states developed their National Reform Programme. For more information, see http://ec.europa.eu/europe2020/europe-2020-in-your-country/italia/national-reform-programme/index_en.htm.

4. For more information on the Estonian Electromobility Program (ELMO) launched in 2011, see http://elmo.ee/en.

References

Appelt, S. and F. Galindo-Rueda (2016), "Measuring the link between public procurement and innovation", *OECD Science, Technology and Industry Working Papers*, No. 2016/03, OECD Publishing, Paris, http://dx.doi.org/10.1787/5jlvc7sl1w7h-en.

Ayyagari, M., A. Demirgüç-Kunt and V. Maksimovic (2011), "Small vs. young firms across the world – Contribution to employment, job creation, and growth", *Policy Research Working Paper*, 5631, The World Bank Development Research Group, http://documents.worldbank.org/curated/en/478851468161354807/pdf/WPS5631.pdf.

European Commission (2016a), "Europe 2020", http://ec.europa.eu/europe2020/index_en.htm.

European Commission (2016b), "Circular Economy", http://ec.europa.eu/environment/circular-economy/index_en.htm.

European Commission (2011), "The Eco-innovation Action Plan", http://eur-lex.europa.eu/LexUriServ/LexUriServ.do?uri=COM:2011:0899:FIN:EN:PDF.

European Commission (2007), "Pre-commercial procurement: Driving innovation to ensure high quality public services in Europe", COM (2007)799 final, http://ec.europa.eu/invest-in-research/pdf/download_en/com_2007_799.pdf.

OECD (2017), "OECD Survey on Strategic Procurement for innovation 2015", in *Public Procurement for Innovation: Good Practices and Strategies*, Annex C, OECD Publishing, Paris.

OECD (2015), *Achieving Public Sector Agility at Times of Fiscal Consolidation*, OECD Publishing, Paris, http://dx.doi.org/10.1787/9789264206267-en.

Chapter 3

OECD Framework to Promote the Strategic Use of Public Procurement for Innovation

This chapter introduces a framework to promote the strategic use of public procurement for innovation. It identifies nine key areas for action to help countries promote the use of procurement for innovation. The framework is based on the OECD Survey on Strategic Procurement for innovation 2015 and the 2015 OECD Recommendation of the Council on Public Procurement.

Introduction

The use of public procurement to boost innovation is still a new policy field in the public sector. The analysis in the previous chapter highlighted achievements some countries have made, and showcased the variety of approaches to procurement for innovation strategies, policies, leadership, implementation, and monitoring that the countries have developed to date. However, often, countries struggle to create an environment and a culture that nurtures innovation. Risk aversion and change resistance among public procurement officials hinder creative solutions. Often, the legal framework for public procurement is highly fragmented, as reported by countries that responded to the OECD Survey on Strategic Procurement for innovation 2015 (hereafter, the "OECD Survey"). Formal requirements are often viewed as additional administrative burden and are therefore eschewed from the outset.

Where the general willingness to conduct procurement for innovation exists, capacity and capabilities are often lacking. To a large extent, public purchasers refrain from using the strategic approach of procurement for innovation to further develop innovation. Although relevant material or sources on public procurement are available on the Internet or are provided by the European Commission, the OECD, the United Nations, the World Trade Organization, stakeholder platforms and others, many procurement authorities are not yet using procurement instruments to encourage innovation and/or have not yet integrated procurement for innovation into strategic policies.

As this report shows, however, many countries have already successfully tackled parts of the issues and countries can learn from these successes. Many countries have undertaken first steps towards procurement for innovation. Strategic procurement for innovation has been applied in areas like energy, environment, food, health, information and communication technology (ICT), transport and water.

This report has analysed a range of good practices and these success stories in procurement for innovation highlighted a set of success factors. This chapter systematises the success factors and proposes a framework for promoting the strategic use of public procurement for innovation.

The framework is designed as a modular and flexible structure and can be applied in a variety of circumstances and levels of governance: at national and sub-national levels and across sectors. In addition, the results of the OECD Survey present new ways of inter-organisational collaboration, new forms of co-operation between governmental and non-governmental actors and their responsible partnerships to create innovative environments. Moreover, the framework illustrates key requirements as fundamental elements of good public governance, to integrate the strategic innovative approach in public procurement, and highlights areas for action from a new perspective to improve co-ordination, governance, management and communication, among other elements required for success.

Main elements

The OECD Framework to Promote the Strategic Use of Public Procurement for Innovation (hereafter referred to as the "OECD Framework") consists of two parts:

- a set of principles (based on the OECD Recommendation of the Council on Public Procurement [hereafter, the "OECD Recommendation"]) that should be followed when planning and implementing measures in support of procurement for innovation
- a mapping of possible measures that can facilitate procurement for innovation.

The OECD Framework covers the entire policy cycle related to procurement for innovation (see Figure 3.1). This illustration is based on the analysis found in Chapter 2, as well as good practices submitted by OECD member countries and non-member economies. Procurement for innovation should begin with the preparation of an action plan or strategy, setting the guideposts for public procurement to implement procurement for innovation. Implementation of the foreseen measures should be followed by evaluation and assessment exercises. Ideally, successful cases of procurement for innovation generate new standard practices. Evaluation and assessment exercises close the feedback loop: they spark changes to the overall procurement for innovation strategy or plan.

Figure 3.1. **Policy cycle in procurement for innovation**

Principles of the OECD Recommendation

As mentioned above, the OECD Framework is based on the OECD Recommendation, and in particular on its twelve integrated principles: transparency, integrity, access, balance, participation, efficiency, e-procurement, capacity, evaluation, risk management, accountability and integration. The principles of transparency, integrity, efficiency, accountability and integration are basic requirements for procurement. The blue-framed principles found in Figure 3.2 are those that are particularly relevant for strategic procurement for innovation: balance, access, participation, capacity, evaluation, risk management and e-procurement (OECD, 2015a).

Figure 3.2. **The 12 integrated principles of the OECD Recommendation of the Council on Public Procurement**

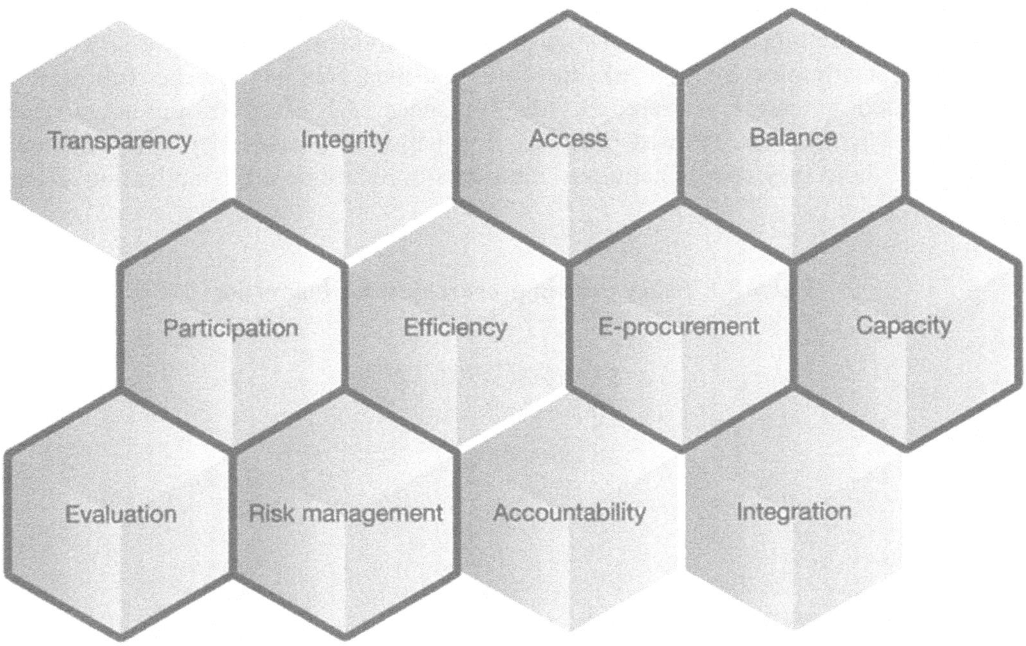

Note: Fields highlighted in blue frames represent the principles that are particularly relevant for strategic procurement for innovation.

1. Balance

One of the recommendations in the OECD Recommendation of the Council on Public Procurement addresses the role of secondary policy objectives. . The "Balance" principle highlights that a well-designed system can also contribute to achieving pressing policy goals but should be **balanced** against the primary policy objective to achieve value for money. These goals include environmental protection, innovation, job creation and the development of small and medium-sized enterprises (SMEs), among others.

"Balance" summarises therefore the need to balance traditional goals of public procurement with secondary policy objectives, such as encouraging innovation. In fact, procurement as a strategic tool for good governance is one of the core issues promoted by the OECD Recommendation. This is particularly relevant when designing strategies to promote innovation.

According to the OECD Recommendation, countries should:

- consider innovation as a secondary policy objective of public procurement and balance it against the primary procurement objectives
- develop a strategy to encourage procurement for innovation
- assess the impact of their procurement for innovation projects.

While all of the 12 integrated principles in the OECD Recommendation are generally important for sound public procurement systems, 6 of them, in addition to, and in connection with, the Balance principle, can be leveraged specifically to support procurement for innovation: access, participation, capacity, evaluation, risk management and e-procurement. The OECD Framework "translates" the principles of the OECD Recommendation for an procurement for innovation context. **Countries aiming to encourage procurement for innovation should pay particular attention to these key principles and focus their efforts accordingly.**

2. Access

One finding in this report is that access for all – and in particular SMEs - is essential to conducting procurement for innovation. Often, rather than established companies, creative newcomers are the source of innovative solutions. While the former often have established access to public procurement opportunities, the latter often are not used to participating in bids or tenders.

The OECD recommends that adherents ensure that companies of all sizes can participate in bids. This is essential for countries wishing to encourage innovation, because red tape presents barriers particularly for small companies that might supply an innovative solution.

Countries should:

- eliminate red tape
- support SMEs in accessing public procurement processes
- keep eligibility requirements and criteria for selecting and assessing suppliers, appropriate

3. Participation

Countries noted the importance of dialogue: both with potential suppliers as well as with the beneficiaries of the innovation. Countries noted that it was in the exchange that the innovation crystallised. It is therefore important to maintain open channels of dialogue and allow for the participation of all relevant stakeholders.

The OECD Recommendation sets out this principle for effective and transparent stakeholder participation. The principle advocates for transparent and regular dialogue, which can be leveraged for innovation generation. The early engagement of suppliers and stakeholders can highlight potential for improvements or flag inadequate solutions, and therefore also act as a risk mitigation measure.

Countries should:

- ensure that dialogues between public officials involved in public procurement and suppliers are organised in a way to generate innovation while remaining fair, open and transparent, such as meet-the-buyer events or supplier seminars for a specific planned purchase
- ensure that all relevant stakeholders are involved in the procurement.

4. Capacity

This principle covers the capacity of the public procurement workforce in terms of sufficient availability of staff skills. Procurement for innovation requires specialised knowledge. Public procurers need to have a high understanding of technical specifications, to make strategic decisions and conduct a relevant and appropriate market analysis. One of the challenges mentioned by countries related to the professional capacity of public procurement staff.

Regular training is particularly important when it comes to supporting procurement for innovation. Similarly, recognition is also key: choosing innovative approaches should not be considered overly risky, but rather encouraged - and successful innovations should be recognised. Knowledge exchange is crucial to encourage innovation as well. Knowledge should not only be exchanged where innovations might originate; the knowledge gained in collaboration with knowledge centres strengthens the capacity of the public procurement workforce to conduct procurement for innovation.

Countries should:

- provide specialised training on procurement for innovation
- have a system in place that rewards innovative solutions
- provide opportunities for staff to exchange with knowledge centres.

5. Evaluation

One argument for embarking on procurement for innovation is that an innovative solution often yields better results than a traditional solution. Without evaluation, however, it will remain unclear whether the innovative solution was indeed better than the traditional path. As the analysis in this report has demonstrated, a minority of countries (44%) currently undertake evaluations related to procurement for innovation.

The OECD recommends evaluating public procurement processes and systems. If countries wish to encourage procurement for innovation, evaluations should be conducted with a view to measuring the effectiveness of innovative approaches. In addition, the OECD Recommendation advocates for the collection of reliable information and data to guide future procurement decisions; this is particularly relevant for procurement for innovation and any decisions concerning whether there might be a better, innovative solution.

A strong evaluation strategy in support of procurement for innovation should include:

- indicators to measure performance of procurement for innovation

- information collection related to the performance of innovative procurement processes
- decision-making processes and needs assessments that leave room for innovative solutions.

6. Risk management

Innovative approaches are usually considered riskier than well-known traditional approaches. Sometimes, innovations might be perceived as riskier, even if they are not. It is therefore important to have an accurate understanding of the risks associated with a new solution (in contrast to the traditional solution), and then take appropriate steps to mitigate risk.

Risk management was addressed by the OECD Recommendation. To facilitate innovation, countries should make sure that this principle is well implemented. They should consider potential specificities of the risk related to innovation in their risk management systems. The role of a sound risk management that addresses potential pitfalls of procurement for innovation will provide an additional source of confidence for public procurement staff.

A sound risk management should also include:

- a higher risk tolerance in procurement for innovation cases
- clear guidelines on how to deal with specific "risky situations" in connection with innovations
- clear and open reporting structures to allow an early response to risks that materialise.

7. E-procurement

The support of public procurement through electronic means represents a major breakthrough in recent years and an innovation in many countries. At the same time, e-procurement introduces many benefits into a public procurement system that can have benefits for innovation as well. In that sense, e-procurement is an enabling factor for many of the other principles mentioned in the OECD Framework.

E-procurement is covered in the OECD Recommendation, which advocates that e-procurement increases access and then competition by simplifying procedures. As noted by countries, innovative companies often have less capacity to participate in public procurement processes. E-procurement can be a lever to encourage their participation. Good e-procurement systems also allow for a flexible response to the developments in an innovative public procurement project. E-procurement systems are scalable; that means they can be applied, for example, both for the pilot and for a broader mass roll-out of a solution. Finally, it is easier to safeguard sensitive information related to innovative solutions in an electronic system.

Countries should ensure that their e-procurement systems serve innovation and innovative companies by:

- keeping the e-procurement system simple and accessible
- maintaining a high standard of confidentiality and security.

Nine areas for action

This report intends to help countries promote strategic procurement for innovation. Procurement for innovation is an opportunity to solve public sector challenges, spanning across borders, ministries and sectors. Consequently, it does not prioritise only "governance"-related issues; it aims to help countries understand a complex system comprised of innovation value chains, which have regional, national and international ramifications.

The principles of the OECD Recommendation that are particularly relevant to the context of strategic procurement for innovation mentioned above provide guidance on how specific actions can encourage procurement for innovation. However, the principles do not prescribe the actions.

Taking into account the challenges and obstacles concerning the development and implementation of strategic procurement for innovation as reported by countries, the following framework is focused on nine areas for action in non-ranking order. The above-mentioned principles can be concretely implemented by choosing actions from these nine categories:

1. policy, strategy and targets
2. legal framework
3. management and leadership in administration
4. financial support
5. professionalisation
6. raising awareness and stakeholder engagement
7. monitoring risks and measurement of impact
8. standards in procurement for innovation
9. e-procurement (as a process-supporting information technology [IT] tool)

These categories represent areas for action that should be present in a sound procurement for innovation framework. Figuratively speaking, measures in these areas bring procurement for innovation to life. The following sections provide good practice cases for each action area. The good practice cases illustrate ways to encourage procurement for innovation and how they should be best implemented. All of the areas are interlinked, while addressing specific parts of the procurement for innovation process.

1. Policy, strategy and targets

National policy strategies to achieve secondary policy objectives differ; regarding procurement for innovation they vary greatly among the countries covered in the present report. The state of play on policies and strategies in countries was discussed and is available in Chapter 2 (see Tables 2.1 through 2.3). Some countries are advanced and have already issued a dedicated procurement for innovation strategy or an action plan, which is in many cases complemented by plans for implementation and impact measurement (see Table 2.1).

Some countries set measurable targets of procurement budgets dedicated to innovation on national level. Other countries are just beginning to address this policy area. Examples of countries with a procurement for innovation action plan as part of the country's general innovation or procurement strategy were listed in Chapter 2 (Table 2.2). In addition, Table 2.3 provided examples of other policy initiatives for procurement for innovation.

Strategic procurement for innovation combines issues that usually fall under the remit of different governmental bodies (e.g. policy making, purchasing, budgeting and scientific research). Therefore, countries face challenges in collaborating across government departments or levels. In addition, it can be challenging to co-ordinate the different responsible bodies as well as other stakeholders to work towards the same goal (see Box 3.1, a spotlight from Spain).

Box 3.1. Spotlight: Innovative practice in Spain

Public Procurement of Innovation Policy in Spain (2010)

The tool consists of different agreements with other public entities to promote their own public procurement actions, benefiting those innovative firms that develop innovations as contracted. The firms that are awarded with the contracts receive not only this contract, but a great impulse for their innovations. Once they sell to the government, the firms have a solid reference that supports their business. Spain mobilised about EUR 200 million in public funds, plus a similar figure of private investments in innovation. In addition, this policy creates strong private-public links. This is a new policy with strong political support (the ministers' council endorsed the policy in 2010). The European structural (regional) funds (FEDER) supported the implementation of this policy.

For more information, see: www.idi.mineco.gob.es/portal/site/MICINN/menuitem.7eeac5cd345b4f34f09dfd1001432ea0/?vgnextoid=281c12c94d36441 0VgnVCM1000001d04140aRCRD.

Source: Country response to OECD (2017), "OECD Survey on Strategic Procurement for innovation 2015", in *Public Procurement for Innovation: Good Practices and Strategies*, Annex C, OECD Publishing, Paris.

Having an overarching, high-level strategy can facilitate co-ordination and collaboration as it sets out guidelines and standards in a transparent manner and shows political will. Any procurement for innovation strategy should be holistic and consider the strategic use of public procurement as stated in the OECD Recommendation. A policy mix should include supply-side and demand-side instruments to stimulate and articulate public demand for innovative solutions. Based on different country policies, it is evident that policy mixes will differ and that the weights in the balancing process will change over time.

The development of a national innovation action plan or procurement for innovation strategy begins with taking stock of goals at the highest political levels. Often, a procurement-related strategy is linked to general innovation, research or economic development policies, as innovation is being generated in scientific environments or in industrial research and development (R&D) departments. In general, action plans follow from an innovation strategy. The action plan should consider horizontal (inter-institutional, inter-ministerial) and vertical (on all governmental levels) teams and working groups.

Strong political commitment is key to the success of any innovation strategy. When present, political commitment and leadership facilitates behavioural changes that lead to an open, innovative attitude among public procurement staff. As a second step, governments should aim to set standards and provide practical guidance to departments and agencies. Thirdly, governments should work to monitor and evaluate outcomes, and assess macroeconomic benefits and levels of compliance. Once the strategy is created, it is important to disseminate it, raise awareness and train all relevant public officials accordingly (as mentioned in the sections below on professionalisation and raising awareness). To ensure early user engagement to build trust and acceptance, a demonstration zone or test fields should be offered (e.g. Climate Streets, an innovative practice used by some countries: a common street is dedicated to sustainability and transformed into a "sustainable street", thereby testing innovative technologies).

In addition, a more holistic use of public procurement as stated in the OECD Recommendation will help to combine secondary policy objectives with various types of instruments, interaction and synergies between policy instruments.

Combining secondary policy objectives

Countries' approaches to combine secondary policy objectives to support the increased efficiency of public procurement systems were mentioned in Chapter 2. This aspect is taken up again under this area of action of the OECD Framework to underscore the potential to secure greater advantages by combining procurement for innovation with other policy objectives. For example, as reported in the compendium Going green: Best Practices for Sustainable Procurement (OECD, 2015b) countries increasingly recognise that green public procurement (GPP) can be a major driver for innovation, providing industry with incentives for developing environment-friendly works, products and services. This is particularly the case in sectors where public purchasers represent a large share of the market, such as construction, health services or public transport.

Most importantly, the recently adopted Sustainable Development Goals (SDGs) feature procurement as a target under SDG 12: Ensure sustainable consumption and production patterns. Under this goal, target 12.7 highlights the promotion of "public procurement practices that are sustainable, in accordance with national policies and priorities". This target means using public procurement strategically to advance national priorities, in addition to promoting sustainability. Public procurement is also linked to SDG 16, which deals with governance aspects. Target 16.6 calls on countries to develop "effective, accountable and transparent institutions at all levels", which also includes public procurement institutions. In supporting the achievement of the SDGs, OECD has already been anticipating the 2030 Agenda by adapting, tailoring and upgrading existing tools (OECD, 2016a).

Combined secondary policy objectives can be implemented at different levels. In this respect, trends in some countries, such as those in Europe, concern sustainable, green procurement show that combined policy objectives will receive more attention, e.g. eco-innovation and the global circular economy. In December 2015, the European Commission adopted the European Union Action Plan for the Circular Economy to develop a sustainable, low-carbon, resource-efficient and competitive economy (European Commission, 2016a), as well as the 7th Environment Action Programme to 2020 (European Commission, 2016b).

The high profiles of socio-economic and environmental concerns are combined to reach citizens and to prepare cities for the future. A significant amount of countries (e.g. Austria, Belgium, Colombia, Denmark and Finland) mentioned initiatives following an Innovative City or Smart City approach at local level, to improve the general welfare of its citizens.

In addition, the establishment of new initiatives for the development of SMEs, e.g. the support for more internationalisation, will have an influence on the stimulation of innovation and future productivity (OECD, 2016b).

2. Legal framework to support strategic procurement for innovation

The legal framework – including regulations, bylaws, and other binding documents that prescribe the rules of the game – is key to a country's national strategic vision. In addition, in some cases, countries are obliged to follow international policy objectives (OECD, 2016a; UNGA, 2015).

Some OECD countries and non-member countries that are also member states of the European Union were obliged to transpose the 2014 European Public Procurement Directives into national law (European Commission, 2014). The EU Directives newly included "Innovation Partnership" and refers to "Pre-Commercial Procurement (PCP)" and "Public Procurement of Innovative solutions (PPI)". Changes in the procedures to facilitate and simplify the procurement process are also included. The better use of life-cycle costs, which describe all the phases through which a product passes from its design to its marketing, and the discontinuation of its production until its disposal as waste or recycling and replacement, will help public authorities to decide on the procurement of R&D services, e.g. in pre-commercial projects.

Existing guidelines or templates may also help countries manage and facilitate procurement for innovation procedures, such as the World Trade Organization's WTO Agreement on Government Procurement (GPA) (WTO, 2016); the UNCITRAL Model Law on Procurement (UNCITRAL, 2011); the World Bank procurement documents (World Bank, 2016); and guidelines published by the European Commission. All these provide support in creating legal frameworks.

The issue of intellectual property rights (IPR) is a particularly important aspect of procurement for innovation. However, according to the analysis presented in Chapter 2, most countries do not seem to place major focus on addressing IPR when conducting procurement for innovation. When procuring innovation, IPR often might not have been settled. Therefore, agreements on rights of ownership and exploitation should be considered as part of the public procurement case. The assignment of IPR can also be used as a means on how to share risks and benefits of a procurement case.

Given the sometimes extensive prohibitions in the legal framework, for example to safeguard against corruption, legal issues are often seen as a main barrier to taking up an innovative procurement procedure. Lack of adequate capability among public procurers and insufficient specialised knowledge of the law were found to be major obstacles to procurement for innovation. To make things more complex, not only the state-of-the-art, but also available technologies, innovations or market developments are expected to be known in detail by the procurer. New education models and trainings on law in relation to procurement for innovation might be helpful (see the section on professionalisation below.)

The plausibility of the action and records of the procurement for innovation process should be retained and available, not only for internal reviews but for audit purposes and social auditing by interested stakeholders (e. g. end users or citizens) to which a public service is delivered. A legal framework for internal and external control creates transparency and helps controls and audits be sufficiently co-ordinated by monitoring of the public procurement system.

Finally, there is a justified fear of conflicts of interest and infringement on integrity and transparency rules in public procurement when public purchasers or state-owned enterprises (SOE) are in close contact with suppliers to discuss new technologies or tasks related to pre-calculations, leading to the prohibition of the participation of a person or company as bidder in a tender, when having contributed to the development of the tender specifications.

3. Management and leadership

An important conclusion of the analysis is that there is a lack of management and leadership in the organisations designated to support the implementation of strategic procurement for innovation. A lack of coordination and cooperation has been identified in most responding countries as big challenges at all government levels. Where the culture is described as risk-averse and work is routine-oriented, it requires strategic and clear management in support of change and interaction, coordination and cooperation.

Even if policy makers take the lead in terms of innovation policy and strategy by making a strong commitment and demonstrating political will to encourage an open innovative government (see the section on policy, strategy, above), procurement for innovation entails a complex decision-making process by procurement officials and department officials. These public "managers" have to lead the organisation and are responsible for the transition of traditional procurement to strategic procurement for innovation, the procurement process and service delivery. As such, these competent managers or "transformational leaders" (Emery et al., 2016), play a key role in procurement for innovation.

It is also important to make use of good practice cases and lessons learned in developing the strategic use of public procurement for innovation; how countries managed to respond to changing society needs; and how countries managed to reap the return on successful investments by using appropriate business models in the public sector.

Another successful solution has been to install an institution or agency that helps manage risk and disadvantages by intermediation (Edler and Yeow, 2016) between the expectation of purchasers and the pending limitations of the offered innovative solutions. Such innovation "intermediaries" are bringing together buyers and suppliers. Developing innovative capacity also depends on the capability of government to lead a broad ecosystem of relationships (OECD, 2015c).

OECD Survey respondents increasingly recognise the potential of strategic procurement for innovation, but administrations often do not see evidence of the benefit. Apart from a legal framework, public administration has to adopt new approaches for co-ordination and co-operation. When managing the procurement cycle in parallel with the innovation cycle, the following management capabilities should be considered: the ability to manage risk; the ability to allocate responsibilities; knowledge of motivation strategies; flexibility (and agility); the ability to facilitate efficient change management; and knowledge of, and ability to use, communication strategies. The procurer is in most cases interested in a risk-reduced process that leads to cost-efficient solutions. Therefore complex procedures between strategic approaches, service delivery and operational procurement as well between public bodies and suppliers and further market actors are challenges. Successful procurement for innovation processes require:

- early communication
- setting up an innovation-friendly environment
- multi-stakeholder collaboration
- minimising the gap between expectations of results and offered solutions from the supply side to avoid organisational failure
- sharing good practice nationally and internationally
- using adequate measurement methods combined with appropriate IT tools.

Among the main challenges to management and governance, lack of internal training measures, insufficient human and financial resources, inconsistencies in management, lack of independent expert advice and insufficient stakeholder management feature highly (see Chapter 2).

A good practice example concerning management in organisations supporting the implementation of strategic procurement for innovation comes from the region of North-Rhine Westphalia in Germany (see Box 3.2). In the example, an early internal communication between staff of different departments took place and the strategic alignment was backed by the procurement department. The management empowered the staff to take the initiative to participate in a European procurement for innovation PCP-project.

Box 3.2. **Spotlight: Innovative practice in Germany**

THALEA enables intensive care units to improve the care for acutely life-threatened patients via telemedicine and telemonitoring (2014)

Research work in the European PCP-Project THALEA is ongoing. Therefore, statements about the outcomes of the innovation are expectations of the project results by the consortium: a growing body of evidence suggests outcome improvement in Intensive Care Unit (ICU) patients by means of telemedicine. At present, a highly interoperable, manufacturer-independent telemedicine-platform for the detection of ICU patients at increased risk does not exist. Encouraging results in other e-health-projects influenced the decision to use pre-commercial procurement (PCP), in order to provide the best possible solution for THALEA. Clearly identified demand and strategy detected by international ICU experts, consented by multidisciplinary stakeholders (IT experts, excellence cluster e-health, insurance companies and ministries) during a pre-consortium meeting ensures a perfect match of demand, strategy and funding instruments in an early phase of the project. Besides unacceptable high mortality in ICU patients, telemedicine has the ability to mitigate problematic pan-European challenges, such as demographic changes, shortage of ICU professionals and scarcity of financial resources. Bringing market participants and stakeholders (procurers, ICU specialists, IT specialists) in close collaboration, PCP within THALEA will create an appropriate common solution, fulfilling the demands of a telemedicine research framework.

For more information, see www.thalea-pcp.eu/.

Source: Country response to OECD (2017), "OECD Survey on Strategic Procurement for innovation 2015", in *Public Procurement for Innovation: Good Practices and Strategies*, Annex C, OECD Publishing, Paris.

4. Financial support

Financial support has two different roles in procurement for innovation: on the one hand, sufficient funding is a necessary prerequisite for undertaking it; on the other hand, the form of funding can act as an important policy lever.

As detailed in Chapter 2, finances can represent one of the main challenges in the procurement for innovation process. Availability of sufficient funds is one aspect. Countries also mentioned fragmentation of funding: in some cases, funding might be available in the budget of one body, but not accessible for another.

The development of innovation bears an imminent risk of failure and loss of financial investment. To encourage the acceptance of these risks, financial incentives are an appropriate instrument to support procurement for innovation in the field of research and innovation. These financial incentives should be aligned with policy strategies and budgets.

In some countries, governments offer different types of financial support within their policy strategies and budget lines: the Small Business Research Initiative (SBRI) programme in the United Kingdom; the Small and Medium Business Administration (SMBA) programme in Korea; or Pre-Commercial Procurement (PCP) and Public Procurement of Innovative solutions (PPI) in the European Union.

Other forms of financial support include tax reduction, flat fees, preferred loan conditions, incentives for procurers, awards or prizes and purchase guarantees.

Box 2.10 on countries' budget targets in Chapter 2 provided examples of qualitative and quantitative targets on innovation or explicit targets on procurement for innovation, depending on the policy, strategy and programme. Table 3.1 provides additional examples of financial support for procurement for innovation activities on national or sub-national levels.

Table 3.1. **Examples of financial support for procurement for innovation activities**

Austria	Financial instruments in Austria are the PCP Programme (awarding of grants to public authorities for pre-commercial procurement) and the PPI Competition (awarding of vouchers on the basis of a contest, which can be used by public procurers for PPI support, such as technology consulting, legal advice, or project management).
Colombia	Financial instruments exist at the national level and sub-national level, provided by the Administrative Department of Science, Technology, and Innovation (*Colciencias*) and the Industry, Business and Tourism Ministry (MinCIT), to mention two of the most relevant.
Denmark	Offers financial support to pre-commercial procurement initiatives (through *Markedsmodningsfonden* and concrete projects related to the government's 2012 innovation strategy).
Estonia	Enterprise Estonia implemented and conducted a pilot programme of respective financial support measures in 2016. Estonia has started some awareness-raising activities, such as hosting the Conference of Innovative Procurements in April 2016.
Finland	Tekes financing instrument for innovative procurement: Since 2009 Tekes has provided funding for innovative public procurement. Funding is typically 50% grant focused to cover additional costs for public procurers in pre-commercial costs, e.g. required expert resources and market negotiation facilitation.
Greece	Greece's smart specialisation strategy (RIS 3) 2014-20 includes a programme on pre-commercial procurement, conducted by the General Secretariat for Research and Technology (GSRT) and the Ministry of Education, Research and Religious Affairs. The programme has a budget of EUR 40 million. A pilot is under preparation.

Table 3.1. **Examples of financial support for procurement for innovation activities** *(continued)*

Spain	An example of specific actions to support procurement for innovation at the national level is the INNODEMANDA programme. Funds will be given to companies to cover innovation costs so that the public body pays the same amount as if it bought the already developed technology, providing greater exposure for these companies' innovative products and services in the administration. Another example is the INNOCOMPRA programme, implemented through FID (Fostering Innovation through Demand) Agreements. This programme uses EU Structural Funds, ERDF, to co-finance procurement for innovations at regional level. Until July 2014, 21 operations had been covered by this instrument, mobilising EUR 230 million.
Sweden	In Sweden, two examples of financial support exist. First launched by VINNOVA, a programme to finance procurement for innovations in 2011. The aim was to encourage Swedish contracting authorities to carry out procurements of innovation and gain experience. Up until 2015, some 35 projects have been financed, mostly pre-studies and pre-commercial procurements, and knowledge about procurement of innovation in Sweden has increased. In addition the catalytic procurements of the Swedish Energy Agency have a history that date back to early 1990. The Energy Agency does not perform procurements itself, but finances and facilitates buyers groups with common needs within specific areas (for example, owners of housing or office buildings). Buyers groups can be comprised of both contracting authorities and private companies.
Switzerland	The National Research Programmes' (NRP) mission is to generate scientific knowledge aimed at solving Switzerland's most pressing problems. The Federal Council specifies the research topics of the individual programmes. NRPs contribute scientifically to the solution of these problems, for example by developing action plans, providing political advice and creating special research infrastructures. The Federal Council usually commissions two to four NRPs at a time with a budget of CHF 10-15 million per project.

Source: Country responses to OECD (2017), "OECD Survey on Strategic Procurement for innovation 2015", in *Public Procurement for Innovation: Good Practices and Strategies*, Annex C, OECD Publishing, Paris.

The European Commission supports the development of procurement for innovation with financial incentives, e.g. in the European Research and Innovation Programme - Horizon 2020 (European Commission, 2016c), and the European Structural and Investment Funds (ESIF) (European Commission, 2016d), by enabling synergies (European Commission, 2016e) between ESIF and Horizon 2020 and other research, innovation and competitiveness-related EU programmes.

5. Professionalisation

Sufficient and highly-trained human resources are central for innovation, and smart public procurement, in general. An adequate capability of procurement officials, as well as professionalisation and awareness related to procurement for innovation contribute substantially to its success. Countries reported that procurement for innovation knowledge, training and advice for the procurement workforce are important elements of successfully using procurement to achieve benefits, but are also seen as challenges. However, collaboration with central purchasing bodies (CPBs), national contact points (NCPs), chambers of commerce and industry (CCIs) and competence centres for procurement for innovation should be developed further to establish a lasting culture of innovation.

Some countries provide good examples of innovation-oriented training that is functioning successfully today. In France, there is a new strategy to develop specific training on procurement for innovation and some countries, including Austria, Belgian (Flanders), Germany and Switzerland, have set up competence centres for procurement for innovation. The aim of these competence centres is to increase the focus on public procurement for innovation.

Feedback loops among public sector purchasers at all levels of administration involved are essential and needed for a systemic approach. In the context of the competence centre, public agencies and procurement bodies should see their potential for innovation stronger than before. Structural, organisational and legal obstacles can be overcome in order to increase the proportion of innovations within an administration.

Apart from these efforts in OECD countries, several initiatives have been taken at the international level: The United Nations Development Programme (UNDP, 2016) offers specialised procurement training on procurement strategy development. The European Commission offers trainings and provides local assistance to public procurers who are interested in implementing procurement for innovations across the EU member states. Two main projects supported by the European Commission are the co-ordination and support action "European Assistance for Procurement for innovation" (European Commission, 2016f) and the "Procurement for innovation Platform" (European Commission, 2016g). In addition, training on e-procurement tools and platforms, "Governance and Capacity Building" (European Commission, forthcoming) should be mentioned as the new EU Directives are driving mandatory practices in all member states.

The OECD developed a toolbox and a checklist for public procurement to support governments in implementing the OECD Procurement Principles (2015a). The toolbox is intended to support policy makers and procurement practitioners at both national and sub-national levels of government (OECD, forthcoming a). In addition, OECD developed a "Roadmap on How to Elaborate a Procurement Capacity Strategy", which includes a sample template for a strategic capacity-building action plan (OECD, forthcoming b). While this action plan was not prepared for the specific context of procurement for innovation, it can be adapted to increase professional capacity to support procurement for innovation (see Annex B).

Chile was able to achieve substantial impact by synergising complementary expertise for the professionalisation of procurement management (see Box 3.3).

Box 3.3. Spotlight: Innovative practice in Chile

Workshop to Improve Regional Public Procurement (2015)

The initiative consists in the joint work of two institutions with complementary expertise, to contribute to the public procurement management of municipalities. ChileCompra provides technical and practical knowledge of public procurement. In addition, the project counts on the participation of the Comptroller General of the Republic (through its regional conptrollers), who provides legal and administrative knowledge. This joint effort is part of the advisory role that the two institutions have among their functions.

Both institutions developed practical workshops. In each municipality, procurement officials are invited. The purpose of the workshops is to involve all relevant actors in municipal procurement, highlighting the fact that public procurement is an institutional activity, where all efforts are relevant (e.g. the requesting area has to provide quality information about its needs, within suitable time periods). In this way, solutions to everyday problems in public procurement can be found at the local level.

Box 3.3. Spotlight: Innovative practice in Chile *(continued)*

This initiative is innovative for the following reasons:

- For the first time, two institutions with complementary expertise in procurement management are working together. This synergy implies better quality workshops, where participants can answer questions more accurately.
- Participants in the workshops included as stakeholders: 1) the requesting areas; and 2) municipal officers. Experience indicates that, in general, procurement officers know the rules of public procurement, however, the requests of other areas are often carried out with inadequate time, or do not provide enough information for the procurement area.

Source: Country response to OECD (2017), "OECD Survey on Strategic Procurement for innovation 2015", in *Public Procurement for Innovation: Good Practices and Strategies*, Annex C, OECD Publishing, Paris.

6. Raising awareness and stakeholder engagement

Communication issues presented major obstacles to procurement for innovation, according to OECD Survey respondents. There are two main aspects to this issue. On the one hand, informing staff and external partners about the benefits of procurement for innovation is important. On the other hand, innovations are often triggered by dialogues between the public sector, companies, (end-)users and other stakeholders. Therefore, there are two types of activities that should be reflected in a comprehensive procurement for innovation policy: awareness about procurement for innovation, and (early) stakeholder engagement.

As explained in Chapter 2, countries should increasingly focus on disseminating the benefits of procurement for innovation to procurement officials as well as to the general public. The countries' responses and collection of good practices highlight how awareness-raising techniques can support culture change and improve attitudes towards procurement for innovation. Secondly, maintaining dialogue and open exchange with partners in an accountable manner can signal areas for change. Procurement institutions should be open to input from outsiders, and frequently seek feedback from (end-)users and suppliers.

Dedicated webpages (European Commission 2016g) are the basis for engaging the public – both for disseminating information about tenders as well as for the dissemination of results. In addition, websites can be a platform to enable exchange and (open) dialogue. In addition, focussed personal exchanges can be organised, for example in the form of workshops or seminars. Interviews and opportunities for comments should be conducted in the planning phase.

Raising awareness by conducting roadshows across countries are tools employed by Austria and New Zealand to reach a broad audience. Thereby procurement for innovation cases are presented, such as described in Boxes 3.4 and 3.5.

Box 3.4. **Spotlight: Innovative practice in Austria (2)**

Mobile traffic management systems - MOVE BEST and MOVEBAG (2011)

The most innovative change here is to have a traffic detection and traffic management system that is mobile and can be used temporarily, wherever and whenever necessary. The objective of the feasibility study, MOVE BEST, was the conceptual design of a mobile and quickly deployable traffic management system for traffic at roadworks and major events. MOVE BEST should sense, analyse and indicate traffic conditions and thereby enable dynamic control of the situation. Unique properties of the system are its rapid availability due to its modular construction using pre-assembled elements; its self-sustaining energy supply; and its wireless data transmission by radio/GSM/UMTS. Well-defined interfaces to traffic control centres also enable the transfer of information to downstream information service providers (radio, web, apps).

MOVEBAG is a flexible, user-friendly, energy-efficient and cost-efficient, safe and easy-to understand (for road users) mobile traffic management system. On the technical side, the system relies upon the component planning tools "*Sensorik, Anzeige and Leitstand*". The conception of these component programmes and their integration into MOVEBAG to create a complete system take place under the correct conditions on an economic, legal, and institutional basis. During the evaluation phase of the system, the feasibility of the concept was established. Out of the seven project proposals submitted, five were awarded the financing of a feasibility study for their idea. On the basis of the first R&D results, two of the consortia got the chance to further work out their idea and develop a prototype of the system. The PCP projects ended with real site testing on the ASFINAG network, based on which ASFINAG is now considering the commercial procurement of the mobile traffic management systems as an innovative new facility to support its everyday work in traffic management on the road.

For more information, see MOVE BEST: www2.ffg.at/verkehr/projekte.php?id=903&lang=de&browse=programm and MOVEBAG: www2.ffg.at/verkehr/projekte.php?id=901&lang=de&browse=programm.

Source: Country response to OECD (2017), "OECD Survey on Strategic Procurement for innovation 2015", in *Public Procurement for Innovation: Good Practices and Strategies*, Annex C, OECD Publishing, Paris.

The second example on raising awareness comes from New Zealand, on a marketing campaign.

Box 3.5. Spotlight: Innovative practice in New Zealand

App4IR crowd sourcing (2014)

This practice relates to a competition to design a new mobile application for Inland Revenue. New Zealand's approach to market was coupled with a marketing campaign and the input of an innovative start-up organisation, Creative hq (*http://creativehq.co.nz/*). This meant that the vast majority of the respondents were either individuals or groups of two to three people. Once the competition was completed, the responses were evaluated by a team of experts and then the short-listed responses were given the opportunity to present their idea to a "dragons den" that consisted of the Chief Executive of Inland Revenue and market leaders from the commercial sector.

For more information, see www.app4ir.ird.govt.nz/.

Source: Country response to OECD (2017), "OECD Survey on Strategic Procurement for innovation 2015", in *Public Procurement for Innovation: Good Practices and Strategies*, Annex C, OECD Publishing, Paris.

7. Monitoring risk and measuring impact

Monitoring, evaluation and managing risks are activities that are part of the public procurement cycle related to procurement for innovation. Only 44% of responding countries have systems in place to measure the impact of their procurement for innovation activities. Yet, monitoring is crucial for tracking and demonstrating the benefits of innovative solutions. Some responding countries have begun to tackle the issue of evaluation and impact measurement by performing studies on a case-by-case basis. The compatibility of different monitoring systems and the adequate use of data and indicators could be more efficient. In addition, sourcing, exploiting and sharing data and information of good practice on models for measurement and impact assessment could contribute not only to procurement for innovation activities, but supports all policies and programmes (OECD, 2015c).

Risk management is another central activity for facilitating procurement for innovation. Agencies perceive procurement for innovation as a riskier-than-traditional form of public procurement. At the same time, resources for assessing and mitigating risk are scarce. Sound risk management systems can achieve two things: they help reduce loss or damage, and they increase trust, because risks related to a process become more transparent and graspable. Suppliers in the bid process normally tend to minimise remaining financial risks for their investments in innovative services or products using clearly formulated risk-sharing arrangements (funding, IPR, guarantees).

Table 3.2 maps activities related to impact and risk management according to the different phases of the procurement for innovation process and the party undertaking each activity. This table can guide the design of monitoring exercises. Some countries successfully measure the impact of their programmes (e.g. Finland and the United Kingdom).

Table 3.2. **Mapping the actions of procurement for innovation phases related to risk management**

	Risk management		
	Preparation	Implementation	Evaluation
Procurers	• Calculate risks • Estimate life-cycle costs • Use consultant expertise • Perform market consultation • Offer a win-win-situation • Offer framework agreements and functional specifications	• Monitor entire process • In case use exit strategies • Use risk management tools	• Perform impact assessment with, e.g. IT tool support • Disseminate lessons learned and benefits of the innovative process • Tweak future process based on evaluation results
Suppliers	• Build trust • Name specifications • Reflect risk-benefit-balance and seek win-win-situation	• Monitor own processes • Protect own IPR	• Exploit/access market • Enhance innovation capacity • Use scaling effects and follow-up projects
End-users	• Build trust • Accept offers for involvement at early stage	• Give feedback at any stage • Test prototypes, use test beds	• Evaluate benefit • Give feedback regarding experience

An example from the Russian Federation showcases impact measurement based on final results (see Box 3.6).

Box 3.6. **Spotlight: Innovative practice in the Russian Federation**

Development of new contracts (2014)

Contracts that are paid for based on their final results: suppliers are paid for the final results that they achieved. These contracts are not the same as well-known Performance Based Contracts or Reportable Test Contracts (healthcare). No payments are made for amount of goods, services or work. The payments are made according to clear and measured indicators of the final effects, which are stated in the contract.

For more information, see http://e-torgi.ru/index.php/stati-2/8002-konets-vsemu-delu-venets.

Source: Country response to OECD (2017), "OECD Survey on Strategic Procurement for innovation 2015", in *Public Procurement for Innovation: Good Practices and Strategies*, Annex C, OECD Publishing, Paris.

In addition in Turkey the Ministry of Development monitors every six months the progress through "Programme Monitoring Reports", see Box 3.7.

Box 3.7. Experience on measurement in Turkey

The Tenth Development Plan, with a strategic perspective at its core, covers economic, social, sectoral and regional areas, as well as setting forth the critical priority areas of intervention through its priority transformation programmes. Programmes are composed of public policies for priority areas that cover more than one sector and facilitate monitoring and implementation of plans. Programme details, sub-components, implementation activities and projects, budget requirements and legislative infrastructure have been transformed into action plans with joint participation and contribution of co-ordinator and responsible institutions for the components. The Ministry of Development monitors the progress through "Programme Monitoring Reports" which are due every 6 months. The High Planning Council is the authority with the right of revision of the programmes if needed, considering the implementation results.

The impact of the "Programme for Technology Development and Domestic Production Through Public Procurement" is monitored by performance indicators, which are defined during the development stage.

Source: Country response to OECD (2017), "OECD Survey on Strategic Procurement for innovation 2015", in *Public Procurement for Innovation: Good Practices and Strategies*, Annex C, OECD Publishing, Paris.

8. Standards in procurement for innovation

Standardisation is often seen as a contradiction to innovation, but it can serve as a catalyst for innovation, especially by defining test standards, methods and quality certificates. In light of globalisation and international trade, the importance of standardisation is evident for national procurement strategies. It also affects different aspects of the procurement process itself (standardised tender specifications, e-procurement, life-cycle costs, production quality, etc.).

The International Standards Organisation (ISO) defines a standard as a document that names "requirements, specifications, guidelines or characteristics" (ISO, 2016) that are prerequisite for ensuring adequate quality in a variety of outcomes, such as goods, services, or processes. Commonly, standardisation refers to the process by which these standards are defined or achieved. The major international standard-setting organisations with regard to procurement for innovation are the ISO or the World Trade Organization (WTO); tradition or the market also influences the formation of standards.

Standards and standardisation are necessary elements of an innovative procurement process for comparability and quality control, from tender specifications to impact assessment and commercialisation of the results. Italy, for example, successfully used an energy efficiency standard to encourage innovative solutions (see Box 3.8). For consumers, standards increase compatibility and interoperability between products. Concerning new technologies, "certified" standards help consumers accept them.

The European Union maintains a reference tool called eCERTIS (European Commission, 2016h), which provides information on certificates frequently required in procurement procedures. It is accessible to any company that wishes to participate in a public procurement procedure. Another European e-procurement solution tool is e-PRIOR (European Commission, 2016i), which is an open-source e-procurement platform. This platform refers to the Pan-European Public Procurement Online project (PEPPOL, 2016), which aims to solve interoperability issues for e-procurement.

Box 3.8. **Spotlight: Innovative practice in Italy**

Integrated Energy Service Framework Agreement 3 (IESFA3)

The Integrated Energy Service Framework Agreement (IESFA) is a performance-based contract for the management of heating, cooling and electrical systems through which the supplier has to guarantee a pre-determined "comfort situation", operation and maintenance, energy savings (measured by means of a measurement and verification programme) and carbon dioxide reduction. Consip's contribution to sustainable market development for energy services started in 2006, when IESFA was published, under the form of Consip's first Energy Performance Contract (EPC). This is a contract in which ESCOs (energy service companies) are motivated and encouraged to optimise energy consumption and resource management in order to improve their profitability and to refund the cost of the energy efficiency measures put in place.

IESFA is currently at its third edition (launched in 2012). The first edition was launched in 2006, the second in 2009. In each edition, Consip introduces more and more challenging energy-saving goals. The tendering procedure has always been an open tender and the awarding criteria has always been the MEAT (most economical advantageous tender), whereby 60% was allocated to price and 40% to quality. The main Energy Efficiency Innovative Solutions, between the second and the third editions, are the following:

- More efficient public services. In addition to heating, suppliers must also achieve electrical savings. This is a "Shared Saving Approach" between supplier and contracting authorities (CAs) that includes an annual flat fee for CAs without any upfront investment or capital expenditure, resulting in guaranteed energy savings for CAs and energy savings for suppliers (ESCOs) as investment recovery offsets investment risk transfer from CAs to suppliers. CAs will own renovated buildings and pay lower O&M+Energy bills.
- Enabling the procurement of innovative products and services. In previous editions, the energy efficiency results were certified by the Italian Electrical Energy and Gas Authority. As part of the IESFA, ESCOs are asked to implement an M&V (measurement and verification) system of energy consumption and savings, as an additional award criterion.

For more information, see www.consip.it/gare/bandi/storico_gare/2012/gara_0018/index.html.

Source: Country response to OECD (2017), "OECD Survey on Strategic Procurement for innovation 2015", in *Public Procurement for Innovation: Good Practices and Strategies*, Annex C, OECD Publishing, Paris.

9. E-procurement

E-procurement represents an important IT tool among others to support the aims of strategic procurement for innovation in the procurement process. The new OECD Recommendation includes the principle of e-procurement to make use of integrated e-procurement solutions covering the public procurement cycle, and to improve efficiency by standardising the procurement process. The degree to which e-procurement is used varies according to a number of factors, including legislation, technology available and the suitability of an electronic process to the particular stage or subject matter of the procurement process (OECD, 2011).

On an individual level, e-procurement involves the obligation of purchasers to explain their decisions, share information to promote transparency, and take responsibility for their actions. The use of appropriate IT tools helps to carry out a proper risk calculation (*ex ante*) and risk assessment (*ex post*). In addition, e-procurement secures the controllability of the complex procurement process (integrity control, combatting fraud, ensuring principles of non-discrimination), and offers the potential to fight corruption through data mining. In times of increasing digitalisation, e-procurement is an essential component of public procurement systems. The use of e-procurement will be mandatory for all European member countries from 2018 on. Some countries have already been pioneers in developing sufficient strategies on e-procurement (see the example from Portugal in this regard in Box 3.9).

Box 3.9. Spotlight: Innovative practice in Portugal

The implementation of e-platforms, most notably the platform called "BASE", can be considered one of the most successful practices implemented to encourage procurement for innovation in Portugal. The use of electronic procurement created an infrastructure that can be considered PPI-friendly.

While Portugal does not have a dedicated legal instrument for procurement for innovation, the most important regulation in Portugal's legal framework related to procurement, the Public Contracts Code (2008), offers instruments to support innovation. This code makes e-procurement mandatory and is in this regard a motor for innovation. As a result of implementing e-procurement, SMEs (either alone or as part of an association) have better access to public markets because tender submission is easier. The code also sets the "most economically advantageous tender (MEAT)" criterion, which enables the contracting authority to consider criteria that reflect technical, innovative and sustainable aspects in addition to price. To facilitate access by SMEs, the code provides for measures like division into lots, adoption of regional criteria, and multi-access criteria.

Source: Country response to OECD (2017), "OECD Survey on Strategic Procurement for innovation 2015", in *Public Procurement for Innovation: Good Practices and Strategies*, Annex C, OECD Publishing, Paris.

Even though countries do not only use e-procurement as an innovative tool for the administration of the procurement process, it remains challenging to offer digital functionalities, such as a dialogue with suppliers, included in e-procurement systems.

Main considerations in applying the OECD Framework

The OECD Framework enhances the implementation of strategic procurement for innovation at all levels of government and across sectors. The OECD Framework consists of two parts: seven principles from the OECD Recommendation that have a particular relevance for innovation, and nine areas for action. The areas for action were developed on the basis of good practices of strategic procurement for innovation. The OECD Framework will help countries implement public procurement for innovation as a strategic means to stimulate innovation through research and development, encourage the market uptake of innovative products and services, and encourage the attainment of other policy objectives, resulting in the long run in increased productivity.

The OECD Framework's modular design allows for flexible implementation concerning achievable policy strategies and development targets. The implementation of strategic procurement for innovation requires building capabilities in the nine areas for action, following the principles of the OECD Recommendation. This means raising sufficient financial resources, equipping agencies with sufficient staff and training both procurement officials and partners in public procurement.

There are many ways countries can leverage partnerships to spur innovation. Partnerships are particularly important in the planning phase of an procurement for innovation project. Many challenges to procurement for innovation pertain to issues that cannot be solved by financial incentives or via regulatory and legal frameworks alone, such as organisational culture and raising awareness about procurement for innovation. Efforts towards harmonisation and standardisation of strategic policies and rules in public procurement by legal decisions and building consensus among national, regional and local governments will help overcome fragmentation. However, countries should widen their use of policy instruments and expand beyond the realm of regulation.

References

Edler, J.and J. Yeow (2016), "Connecting demand and supply: The role of intermediation in public procurement of innovation", *Research Policy*, Vol. 45, Issue 2, www.sciencedirect.com/science/article/pii/S0048733315001638.

Emery, Y. et al. (2016), "Towards innovative public services: A framework for the development of the innovation capability of European Public Administrations", European Public Administration Network, EUPAN, www.eupan.eu/files/repository/20160219145949_Annexes_-_Towards_Innovative_Public_Services_%28Final_Version%29_-_EUPAN_-_Luxembourgish_Presidency_2015.pdf.

European Commission (forthcoming),"Governance and capacity building", http://ec.europa.eu/growth/single-market/public-procurement/e-procurement/index_en.htm.

European Commission (2016a), "Circular Economy", http://ec.europa.eu/environment/circular-economy/index_en.htm.

European Commission (2016b), "Eco-innovation Action Plan", http://ec.europa.eu/environment/ecoap/.

European Commission (2016c) "Horizon 2020", https://ec.europa.eu/programmes/horizon2020/en/what-horizon-2020.

European Commission (2016d), "European Structural and Investment Funds, ESIF", http://ec.europa.eu/regional_policy/en/information/legislation/regulations/

European Commission (2016e), "Enabling synergies between European Structural and Investment Funds, Horizon 2020 and other research, innovation and competitiveness-related Union programmes - Guidance for policy makers and implementing bodies", http://ec.europa.eu/regional_policy/sources/docgener/guides/synergy/synergies_en.pdf

European Commission (2016f) "European assistance for procurement for innovation", https://ec.europa.eu/digital-single-market/en/news/training-promotion-and-local-implementation-assistance-pcp-and-ppi.

European Commission (2016g), "Innovation Project Procurement Platform", www.innovation-procurement.org and www.innovation-procurement.org/about-ppi/guidance.

European Commission (2016h), "Information system eCERTIS", https://ec.europa.eu/growth/tools-databases/ecertis/.

European Commission (2016i), "E-procurement platform ePRIOR", http://ec.europa.eu/dgs/informatics/supplier_portal/index_en.htm.

European Commission (2014), "Procurement Directive 2014/24/EU, replacing Directive 2004/18/EC", http://eur-lex.europa.eu/legal-content/EN/TXT/?uri=celex:32014L0025 and "Procurement Directive 2014/25/EU, replacing Directive 2004/17/EC (Utilities)", http://eur-lex.europa.eu/legal-content/EN/TXT/?uri=celex:32014L0025.

ISO (International Standards Organisation) (2016), "Standards", webpage, www.iso.org/iso/home/standards.htm.

OECD (forthcoming a), "Toolbox and a checklist for public procurement", OECD.

OECD (forthcoming b), "Roadmap: How to elaborate a procurement capacity strategy", MENA-OECD Network on Public Procurement.

OECD (2017), "OECD Survey on Strategic Procurement for innovation 2015", in *Public Procurement for Innovation: Good Practices and Strategies*, Annex C, OECD Publishing, Paris.

OECD (2016a), "Towards an OECD Strategic Response to the Sustainable Development Goals: Consultation draft", February, www.oecd.org/dac/Consultation-draft--Outline-OECD-strategic-response-to-the-SDGs.pdf.

OECD (2016b), *Financing SMEs and Entrepreneurs 2016: An OECD Scoreboard,* OECD Publishing, Paris, http://dx.doi.org/10.1787/fin_sme_ent-2016-en.

OECD (2015a), "Recommendation of the Council on Public Procurement", www.oecd.org/gov/ethics/OECD-Recommendation-on-Public-Procurement.pdf.

OECD (2015b), Going Green: Best Practices for Sustainable Procurement, www.oecd.org/gov/ethics/Going_Green_Best_Practices_for_Sustainable_Procurement.pdf

OECD (2015c), *The Innovation Imperative in the Public Sector: Setting an Agenda for Action*, OECD Publishing, Paris, http://dx.doi.org/10.1787/9789264236561-en.

OECD (2015d), *The Innovation Imperative: Contributing to Productivity, Growth and Well-Being*, OECD Publishing, Paris, http://dx.doi.org/10.1787/9789264239814-en.

OECD (2011), "E-Procurement", *SIGMA Public Procurement Briefs*, No. 17, OECD Publishing, Paris, http://dx.doi.org/10.1787/5js4vmnvwp8w-en.

PEPPOL (Pan-European Public Procurement Online) (2016), "About PEPPOL", http://peppol.eu/about-openpeppol/.

UNCITRAL (United Nations Commission on International Trade Law) (2011), "Model Law on Procurement, www.uncitral.org/uncitral/en/uncitral_texts/procurement_infrastructure/2011Model.html.

UNDP (United Nations Development Programme) (2016), "Procurement training", www.undp.org/content/undp/en/home/operations/procurement/procurement_training/.

UNGA (United Nations General Assembly) (2015), "Transforming our world: the 2030 Agenda for Sustainable Development", Resolution 70/1 adopted by the General Assembly on 25 September 2015, A/RES/70/1. Goal 12.7., https://sustainabledevelopment.un.org/post2015/transformingourworld.

World Bank (2016), "Procurement for Projects and Programs", http://web.worldbank.org/WBSITE/EXTERNAL/PROJECTS/PROCUREMENT/0,,contentMDK:20062006~menuPK:84284~pagePK:84269~piPK:60001558~theSitePK:84266,00.html.

WTO (World Trade Organization) (2016), "Agreement on Government Procurement", www.wto.org/english/tratop_e/gproc_e/gp_gpa_e.htm.

Annex A

Factsheets from the country responses to the OECD Survey on Strategic Procurement for innovation 2015

Note by Turkey: The information in this document with reference to "Cyprus" relates to the southern part of the Island. There is no single authority representing both Turkish and Greek Cypriot people on the Island. Turkey recognises the Turkish Republic of Northern Cyprus (TRNC). Until a lasting and equitable solution is found within the context of the United Nations, Turkey shall preserve its position concerning the "Cyprus issue".

Note by all the European Union Member States of the OECD and the European Union: The Republic of Cyprus is recognised by all members of the United Nations with the exception of Turkey. The information in this document relates to the area under the effective control of the Government of the Republic of Cyprus.

Austria[*]

Strategic framework, stand-alone action plan and scope for procurement for innovation policy

Austria implemented the "Austrian Action Plan on Public Procurement Promoting Innovation PPPI" as a follow up of the "Austrian Strategy for Research, Technology and Innovation (2011)". The strategy aims to create a "systemic, modern policy on research, technology and innovation" by using public procurement as a lever. The strategy has linkages to other policy areas, for example education policy and competition policy.

The PPPI Action Plan was adopted in 2012 and is based on a 12-month PPPI Strategy Process, involving all relevant Austrian stakeholders (about 100). The responsibility for the PPPI Strategy Process as well as for the implementation of the PPPI Action Plan has been cooperatively supervised by the Ministry for Transport, Innovation and Technology (BMVIT) and the Ministry of Science, Research and Economy (BMWFW), supported by the Austrian Procurement Agency (BBG) and the Austrian Institute of Technology (AIT).

The action plan includes a number of specific actions or initiatives to support procurement for innovation. The most important activities are: 1) the amendment of the Austrian public procurement law; 2) the creation of a PPPI service network consisting of a PPPI service centre (including personal and online services) and PPPI service partners (complementary to the PPPI service centre, by covering sectors such as mobility, energy, building, managing grants, connecting to the industry, facilitating exchange with the Austrian provinces, etc.); 3) the design of an advanced PCP-programme (e.g. transport infrastructure); 4) the management of an innovation platform; 5) the initialising of pilot projects; and 6) the creation of linkages to already existing procurement initiatives, such as the federal Green Public Procurement Action Plan. Another important aspect is the PPPI awareness initiatives and trainings under the responsibility of different ministries and the PPPI service network.

Austria uses the European PCP/PPI definitions. PPPI (Public Procurement Promoting Innovation) is the umbrella term which includes both, PCP and PPI.

Implementation

Austrian good practice policy in PPPI is above all the "empowerment approach" and its institutional backing. That is, the design of various interlinked measures to empower public procurers for procurement for innovation and their establishment in already existing innovation supporting institutions, which is among others reflected in the PPPI service network approach, the PCP-programme, etc. Beyond that, the Austrian "evidence-based policy approach" is good practice in PPPI (monitoring, evaluation).

Austria designed a PCP-programme that requires a financial contribution of the involved public procurers. This minimises or even eliminates the risk of non-procurement after the completion of the PCP process.

[*] OECD Survey Part I submitted by Eva Buchinger, Scientist, Austrian Institute of Technology (AIT) on behalf of the Ministry of Science, Research and Economy (BMWFW) and the Ministry for Transport, Innovation and Technology (BMVIT).

Austrian example cases of good practice procurements are, among others, eco-buildings, LED lights, smart grids, traffic control and risk detection, and e-government. It is part of the overall Austrian good practice that empirical evidence of good practice of public procurement for innovation is collected and distributed in the course of the PPPI platforms.

An procurement for innovation monitoring system has been set up. It includes an procurement for innovation survey by Statistics Austria, monitoring obligations of the PPPI service centre and scientific interim and *ex post* evaluations.

Policy instruments used to support procurement for innovation in Austria are the Procurement for innovation Action Plan that aims at all sectors. Financial instruments are the PCP Programme awarding of grants to public authorities for pre-commercial procurement and the PPI Competition awarding of vouchers on the basis of a contest which can be used by public procurers for PPI-support, such as technology consulting, legal advice, or project management. Considerable success has been achieved until now, especially in the sectors of mobility, energy, buildings and information and communication technology (ICT).

Challenges, risks and solutions to overcome obstacles

The most significant challenge in implementing the PPPI Action Plan is the mobilisation of the procurers. Procurers respond slowly because procurement of innovation initially requires more time (and money) compared to standard procurements. It also requires a procurement strategy together with the involvement and commitment of the (top) management of the procuring organisation, and requires the acceptance of some risk and/or uncertainty (i.e. risk of innovation failure).

To overcome these obstacles, stakeholder participation, procurer empowerment and the offering of service/support have been used in all stages - starting from the strategy process which resulted in the Austrian PPPI Action Plan. Concerning PCP, the biggest challenge was the IPR-agreement between procurer and industry. In the PCP-pilot we managed step-by-step to address procurers and industries needs appropriately and found an acceptable solution for all parties.

Key lessons learned

- Stakeholder participation is one of the key elements for the general acceptance of PPPI policy.
- Embedding individual activities into the national innovation strategy is important, as is overall political commitment for PPPI.
- Empowerment works, but requires time and financial resources.
- An appropriate policy mix is necessary (i.e. there needs to be a broad range of instruments from awareness to programmes and services to financial incentives).
- Adopting an evidence-based approach is important – this requires monitoring and evaluation.

Measurement and impact assessment

There is no quantifying of targets for procurement for innovation. To capture the impact of procurement for innovation activities, the Austrian PPPI Action Plan covers various dimensions: increasing (significantly) the share of innovative procurement in public procurement and increasing (significantly) the share of procurement-oriented research and development (R&D). There are also other indicators such as: reducing environmental burden, reducing costs (within public entities), improving processes (within public entities), and improving public service quality (benefits for citizens). Beyond these dimensions, an procurement for innovation monitoring system has been set up. It comprises an procurement for innovation survey by Statistics Austria, monitoring obligations of the PPPI service centre and scientific interim and ex post evaluations and covering the following types of procurement for innovation: procurement of goods/services newly developed for the procuring entity, first commercial procurement of goods/services and the diffusion of innovative goods/services.

An interim impact assessment took place in 2014. An assessment of the PPPI service centre was conducted in 2015. A comprising impact evaluation will take place in 2017/18.

Belgium*

Strategic framework, action plan and scope for procurement for innovation policy

There is no dedicated procurement for innovation action plan at the federal level, but ad hoc initiatives exist. The federal administration focused on the development of an e-procurement platform, available to all Belgian public administrations. Other elements are: facilitation of SME participation, sustainability, maximising competition in framework agreements.

Recently at federal level a new project just started. The objective is to implement a new procurement process to make public procurement accessible for start-up companies with a minimum of regulation and to quickly implement innovative solutions.

At the regional level, initiatives are taken as well. For example, the Flemish government approved an Action Plan on Procurement of Innovation in 2008. This action plan focuses on procurement for innovation that needs a pre-commercial research and development phase, i.e. pre-commercial procurement. The Flemish Agency for Innovation by Science and Technology (Instituut voor Innovatie door Wetenschap en Technologie – IWT, this organisation has now been absorbed by the newly formed "Flanders Innovation and Entrepreneurship", as from 1 January 2016.) manages the pilot programme under this action plan. It aims at a horizontal integration in the innovation policy mix, whereby the policy domains of government buy innovation from companies and research centres. The government is the first receiver of innovative solutions and the society can be supplied by innovative products in some lead markets. The innovative procurement instrument should become a fully integrated part of a balanced innovation policy mix strategy. The Flemish ministers decided on participation for their respective area and nominated a contact point that identifies the challenges and opportunities that can be solved through procurement for innovation. Through this process, 48 project proposals were received and 15 selected. The first pilot was located in the cultural sector and provided a digital book platform. Other procurements that have been launched include an eye screener for young babies, monitoring systems for excavations, POP (personal development plan), etc.

Wallonia public procurement policy focusses on fostering sustainability and ethical and social clauses, which can be a driver for (eco-) innovation. Dematerialisation of public procurement is another priority, aiming at simplifying and facilitating access of SMEs. These actions are part of the regional development strategy, and are anchored in several action plans, such as the Marshall Plan 2.green, the Walloon Small business Act, the simplification and e-gov action plan, Environment-Employment Alliance, etc. Specific actions on procurement for innovation remain a challenge for Wallonia.

Implementation

Aside from implementing latest EU directives, procurement for innovation at the federal level is conducted via ad hoc initiatives. Examples are the development of an e-

* OECD Survey Part I submitted by Waldo van den Broeck, Adviseur-generaal, Diensthoofd, Overheidsopdrachten, Federale Overheidsdienst Personeel en Organisatie.

procurement platform and guidelines sustainable buying, trainings and workshops on social economy (clothing) related to the preparation of tender documents.

Flanders: The Flemish administration is currently implementing its action plan on procurement for innovation (see above.) Aside from that, the administration is exploring innovation-friendly ways to boost commercial procurement of innovation by the Flemish government, thereby principally stimulating the PPI-path through setting some indicative targets. Certain lead areas provide guidance for improvements related to procurement for innovation, such as Economy, Science and Innovation (Department of Economy, Science and Innovation), Public Governance (Public Procurement Division) and Services for the General Government Policy (Government of Flanders Executive Office Division) Innovation can be stimulated through traditional public procurement by means of eliminating (or at least reducing) legal, financial, precautionary or other hurdles that hinder tender participation of innovative actors (e.g. SME and young innovative companies, etc.).

The Walloon region introduced guidance and standard clauses to facilitate green procurement and SME-friendly procurements.

Brussels Region commenced the Smart Cities Mobility Platform project. This project aims at better integrating ICT tools in the mobility ecosystem of the region, for example through apps, traffic management systems and data mining of mobility data.

Good practices

At federal level, a successful practice was the development of the e-procurement platform led to a single application used by all economical operators and by all administrations.

At the regional level, a good practice was submitted by Flanders: Flanders was one of the first EU regions that launched the first PCP and PPI's projects. IWT's Knowledge Centre on innovative procurement was crucial for the projects' success. The Centre elaborated a viable methodology consisting of two steps, first, to detect and spot suitable innovation projects from procurers, and second, to assess needs of the end-user requirements.

The mapping and fine-tuning of the procurers' needs remain a crucial aspect to fully deploy procurement for innovations; after mapping the needs, it is essential to bridge the gap with the supply-side. Therefore, market consultation sessions are important to match the supply and demand-side. In these sessions, information on the capabilities of suppliers and the feasibility to develop the envisaged technological solution can be retrieved. In addition, these sessions serve to clarify potential risks.

Successful policy instruments

Successful policy instruments include the Smart@Fire project methodology (on regional level). As part of this project, IWT in Flanders developed a methodology that was approved by the European Commission.

At the national level, due to a restrictive interpretation of the procurement legislation, the PCP procurement procedure is still confronted with obstacles to be resolved in order to allow a flexible, timely and more innovation friendly implementation. This corresponds with the envisaged, more PPI-oriented approach mentioned above. In total, seven PCPs have been launched (one is finalised, three are awarded and in the development phase, 3 have been stopped), three PCPs are waiting for approval, five PPIs have been successfully finalised.

A single portal for public procurement, with a toolbox and a helpdesk available to facilitate insertion of ethical, social and environmental clauses in the procurement documents. In that perspective, a facilitator network was also created to sensitise and train procurers at the on regional level.

Challenges, risks and solutions to overcome obstacles

Main challenges are related to obstacles in the national legislation and a restrictive interpretation of the exemption for R&D (PCP -art 16f Procurement Directive), the lack of financial commitment and interest in the subject, the low level of knowledge about innovation and the PPI-PCP procedures, etc. Another major challenge is the lack of personnel capacity appointed for the bottom-up implementation of such a high-level instrument with a broad target group of entities from 13 policy areas.

PCP and PPI in the future.

Key lessons learned

On the level of knowledge there is still a long way to go in order to enhance the awareness, general understanding and willingness to invest in PCP-PPI. Some "Meet and greet" sessions will be organised to establish contact with buyers.

Measurement and impact assessment

Belgium has quantified its 3% target for procurement for innovation and measures/follows up moves to reach of this target. But the target actually has not yet been achieved. For the time being there are no impact assessments, evaluation studies and/or studies of state of play regarding any type of procurement for innovation on national level, but on regional level (Flanders) studies of state of play.

At federal level, the use of the e-procurement platform by all administrations is targeted and measured. Indicators for SME participation are currently being implemented.

In the near future, Flanders' interdepartmental working group on procurement of innovation will implement a monitoring system for the 3% target.

Canada[*]

Strategic framework, action plan and scope for procurement for innovation policy

The federal government of Canada has an Economic Action Plan of which the Build in Canada Innovation Program (BCIP) is part of. Industry Canada has Canada's innovation strategy entitled "Seizing Canada's Movement".

The Action Plan Contracting Policy from the Treasury Board Secretariat (TBS) provides a procurement framework. Public Services and Procurement (formerly PWGSC) derive the procurement policies from the TBS policy and Financial Administration Act aligning with international trade agreements.

The specific action to support procurement for innovation at the national level of government is the BCIP. PWGSC developed a SMART procurement framework which uses four elements: Early Engagement, Effective Governance, Independent Advice and Benefits for Canadians to support innovation in Canada.

PWGSC is also sharing procurement tools with provinces, municipalities and other government agencies which enable all levels of government to meet procurement obligations and leverage good procurement practices.

The BCIP supports innovation in Canada while aligning with the country's international trade agreement obligations and Canadian fair, open, transparent and competitive procurement values.

Challenges, risks and solutions to overcome obstacles

Canada noted that it was a challenge to align national policies with international trade agreements, while creating a fair and competitive procurement process that allows for the purchase of creative, innovative and not-yet-to-market products which have commercial potential. These obstacles have been overcome.

Key lessons learned

Canada stated that a lesson learned was to find a way in which Canada could enhance outreach activities to the benefiting departments to match innovations while building supplier confidence in the process.

Measurement and impact assessment

The BCIP was subject to an internal departmental evaluation as it transitioned from a pilot to a permanent programme. The evaluation examined the programme's activities to assess its relevance and performance. The pilot of the BCIP was found to have a continued relevance and the ability to deliver a unique form of assistance when compared to other programmes.

[*] OECD Survey Part I submitted by Paul Thomson, Manager, Canada Policy, Risk, Integrity and Strategic Management Sector Public Services and Procurement.

The BCIP performs impact assessments/trends analysis on the data obtained from evaluation questionnaires sent to both testing departments and suppliers upon completion of testing and one year post testing. Related performance indicators that are tracked include the number of testing departments using innovations in operations, the percentage of innovations launched into the marketplace and the number of innovations for which additional quantities have been procured beyond testing.

With the collaborative efforts of multiple government organisations and industry partners, the BCIP helps innovators to bridge the pre-commercialisation gap by helping them move their innovations from the lab to the marketplace through testing in operational environments across government. The BCIP awards contracts to entrepreneurs with pre-commercial innovations through an open, transparent, competitive and fair procurement process for their testing within the Canadian federal government. The programme facilitates testing opportunities within the federal government with testing departments being required to provide feedback to entrepreneurs on the performance of their goods or services. In doing so, the BCIP provides innovators with the opportunity to enter the marketplace with a successful application of their new goods and services. With the help of the Office of Small and Medium Enterprise's Network of Public Works and Government Services Canada, the BCIP also provides information on how to do business with the government of Canada.

Chile*

Strategic framework, action plan and scope for procurement for innovation policy

Currently, in Chile there is no strategic framework to manage innovation in public procurement. However, since 2014, the Dirección ChileCompra has been establishing a system for innovation management which focuses on promoting market integrity in public procurement.

Chile does not have an procurement for innovation action plan at the moment.

Currently, there is not an procurement for innovation policy in Chile. However, it seems evident the necessity to move forward in that direction. ChileCompra, is working in establishing capacities between its collaborators, generating networks with other public agencies in order to promote an innovation culture.

Implementation

As part of the general innovation management system for increased integrity, Dirección ChileCompra is developing two projects that are in the implementation phase. One of these projects consists of workshops with municipalities from different regions of the country. The workshops aim to improve the integrity of their procuring processes. This strategy is designed to work in co-ordination with the Regional Comptroller, responsible for auditing the public procurement processes at local level. The other project aims at creating indicators that decision makers in public entities can use to evaluate the procurement performance and make corrective actions as necessary.

Challenges, risks and solutions to overcome obstacles

In Chile's experience, the most significant challenges have been:

1. Resistance to change.
2. Considering failures as an opportunity to learn.
3. Innovation and urgency are difficult to coexist. While innovation needs time to understand and investigate a problem, to share experiences and listen to the client's opinions to create the best solution, institutional goals need to be urgently satisfied.
4. Modification of the regulatory framework.

Key lessons learned

ChileCompra has experience with pursuing secondary policy objects of public procurement (for example sustainable green growth, the development of SMEs, local job creation, etc.). However, Chile has not yet focused on innovation, but is conscious that it would be important to include innovation as part of the secondary policy objects in the near future.

* OECD Survey Part I submitted by Molina Maria de los Angeles, Jefe Observatorio ChileCompra (S), Dirección de Compras y Contratación Pública, Chile.

Colombia*

Strategic framework, action plan and scope for procurement for innovation policy

The National Development Plan (2014-18) specifies procurement innovation as a cross-cutting strategy targeted to generate a higher economic and social value to enhance the conditions for the development of business activities. Procurement innovation is also conceived to provide the demand of good and services for specific needs in a more efficient way through innovation. This part of the governance strategy will be complemented with accompaniment services for the SMEs in the implementation of the necessary processes and investments to achieve higher quality standards.

The procurement for innovation action plan is part of the country's general innovation or procurement strategy. The National Development Plan (2014-18) institutes procurement for innovation as a cross-cutting public policy to increase value for money through procurement of innovative goods and services, and to promote innovation in the Colombian market with a demand-driven approach.

Colombian regulation refers to procurement of science, technology and innovation with no particular emphasis in innovation. Such regulation allows government agencies to contract without competition when the purpose of the contract involves scientific investigation and development projects with innovative solutions.

Implementation

Specific actions: There are three pilots:

- Ministry of ICT (MinTIC): the acquisition of services for the development of a practice lab on IT management and information security.
- National Agency for Overcoming Extreme Poverty (ANSPE): the development and acquisition of an IT-based tool used to strengthen soft skills - among citizens in extreme poverty conditions - that have an impact on income generation initiatives targeted to the same public.
- Empresas Públicas de Medellín (EPM): for the acquisition of a real time water leak detection solution. Additionally, the procurement for innovation policy has been included within the communication plans of the main members of the so-called National System of Science, Technology and Innovation (hereafter "STI System").

In spite of the fact that we believe there are experiences of procurement for innovation in Colombia, there are no records thereof what makes no possible to list the best practices and results up to now. Colombia Compra Eficiente started implementing the procurement for innovation policy on 2015 and results will be available for analysis by the end of 2016.

* OECD Survey Part I submitted by Maria Margarita Zuleta Gonzalez, Director, Colombia Compra Eficiente.

Policy instruments: The National Development Plan highlights procurement for innovation as a high priority policy to be implemented in the 2014-18 period. The regulatory framework for the STI System and related procurement is mentioned above. Innovation promotion initiatives are developed by government agencies, both at the national and sub-national level, mostly supply-driven and focused on possible demand from the private sector.

Financial instruments exist at the national level and sub-national level, provided by the Administrative Department of Science, Technology, and Innovation (Colciencias) and the Industry, Business and Tourism Ministry (MinCIT) to mention two of the most relevant.

Challenges, risks and solutions to overcome obstacles

The current regulatory framework enables the procurement for innovation policy allowing PCP and PPI processes; however its application is a challenge in a risk-averse culture amongst public officials fuelled by tight fiscal controls and cases of corruption in the past. The development of innovation-related competencies in public officers is a key success factor to overcome this fear and improve the results of the procurement process by strengthening the planning phase. On the other hand, there is a communication challenge to explain to the private sector the policy and the instruments developed for procurement for innovation. Quick wins are important and useful for followers. Financial instruments, for buyers and suppliers alike, should be strengthened in order to promote procurement for innovation.

It is too soon to state that we have overcome these obstacles, because the pilots are still in an early phase.

Key lessons learned

Procurement for innovation is not in the mind of public officers when they study the options to procure goods and services. Further, if they consider procurement for innovation the absence of precedents and the fear of a new method make them nervous. On the other hand, private sector is not familiar with procurement for innovation. Therefore, Colombia Compra Eficiente shall prepare training and communication tools addressed to government agencies and suppliers and work with the controlling authorities to offer comfort to buyers and suppliers.

Cyprus*,+

Strategic framework, action plan and scope for procurement for innovation policy

Cyprus does not have a strategic framework for procurement for innovation and no secondary policy objectives have been set for public procurement. However, in accordance with EU public procurement legislation, public procurement practices demand that every contracting authority assesses the needed innovation parameters of the contract concerned. Contracting authorities can do so at their discretion; the innovation parameters are to be reflected in the technical specifications or other procurement documents. Thus, procurement for innovation is conducted on a case-by-case basis. In addition, Cyprus's Employers and Industrialists Federation has established innovation awards for certain sectors, including public sector innovation achievements.

Measurement and impact assessment

There is no system in place to measure the impact of actions related to procurement for innovation. The required innovation level is assessed a-priori (before the contest publication) and becomes a contractual obligation of the contractor.

* OECD Survey Part I submitted by Christos Aspris, Officer, Directorate General for European Programmes, Coordination and Development.

\+ *Note by Turkey*: The information in this document with reference to "Cyprus" relates to the southern part of the Island. There is no single authority representing both Turkish and Greek Cypriot people on the Island. Turkey recognises the Turkish Republic of Northern Cyprus (TRNC). Until a lasting and equitable solution is found within the context of the United Nations, Turkey shall preserve its position concerning the "Cyprus issue".

Note by all the European Union Member States of the OECD and the European Union: The Republic of Cyprus is recognised by all members of the United Nations with the exception of Turkey. The information in this document relates to the area under the effective control of the Government of the Republic of Cyprus.

Czech Republic*

Strategic framework, action plan and scope for procurement for innovation policy

In the Czech Republic there is no single policy document for setting a strategic framework for procurement for innovation (pre-commercial or commercial), and no innovation action plan. Nevertheless, public procurements in R&D and innovation are traditional tools of support to innovation solutions for public sector and an integral part of the Czech innovation policy. A conceptual shift regarding procurement for innovation is envisaged to stem from the key policy document for RDI policy, the "National Research, Development and Innovation Policy of the Czech Republic in 2009-2015 with an outlook to 2020," updated in 2013. Based on this strategic document, there is interest in improving framework conditions (including legal conditions) for financing research and innovation projects of SMEs through public procurement. Furthermore, by 2016 the European Commission and Council Directive for public procurement will be transposed into the Czech legislation.

Currently, the main scope for procurement for innovation is based on several instruments. The procurement of R&D and innovation is based on the Act on Research and Development Support from Public Funds. Every procurement process is part of this programme previously authorised by the government. Most procurements are part of the programme BETA managed by the Technology Agency of the Czech Republic (TACR). The procurement procedure itself is governed by the Act on Public Procurement. The main differences to the commonly used PCP concept are firstly, that the exclusive owner of research results is the sponsor, i.e. governmental or public body, not the inventors themselves, and secondly, that there is no phasing during a project implementation in use. The PCP concept as described by the European Commission constitutes the basis for the Operational Programme Enterprise and Innovation for Competitiveness managed by the Ministry of Industry and Trade.

However, there have been no calls for PCP project proposals so far. TACR will finalise a methodological framework on PCP (see below) in October 2015. It also prepares new a programme that will replace the BETA programme (2012-16) in the future. TACR aims at creating a target that allots one third of the budget of the new programme to PCP. A new type of procurement procedure will be part of the future act on public procurement. The innovation partnership will be transposed from the directive 2014/24/EU on public procurement into Czech legislation. This procurement procedure is very similar to PCP. It is divided into phases (mini tenders); in addition, there are various options regarding IPR ownership. There are no legal definitions on procurement for innovation. Other official documents use the terms PCP or PPI in English or Czech. These are defined similarly to the definitions in EC documents.

* OECD Survey Part I submitted by Lukas Levak, Director of the Department of Research and Development, Ministry of Education, Youth and Sports.

Implementation

There are no specific actions to support procurement for innovation at the national/central/federal level of government. However, the Czech Republic has implemented several programmes designed for the fulfilment of research and innovation needs of public sector. These are considered a good practice:

- BETA programme (2012-2016), implemented by the Technology Agency of the Czech Republic (TACR)
- Programme for defence applied research, experimental development and innovation (2011-17), operated by the Ministry of Defence
- Programme for security research for the needs of the state 2010-15, operated by the Ministry of Interior.

All these programmes apply public procurement as a tool for selection of research and innovation projects directed to solutions for specific needs identified by respective public bodies. However, none of these programmes use the concept of PCP for selection of innovation solutions.

Since 2012, TACR intensively explores possibilities for implementation of the PCP concept within the existing legal framework. In collaboration with the TAFTIE network and Czech public sector institutions TACR gradually formulates a methodological framework for PCP in the Czech Republic. Procurement in the form of PCP was launched as a pilot in 2014, addressing the research needs of the Czech Ministry of Interior. The procurement goal is to create free software for archival description, which will be open to all the archives and other "memory" institutions in the country.

The final version of the methodological framework for PCP will be delivered in October 2015. It will be accessible for everyone and all public authorities will be able to use this methodological framework within contemporary and also future legal framework. It is based on the framework agreement concluded between contracting authority and all interested suppliers according to the act on public procurement. Suppliers will oblige to their participation in the procedure consisting of several phases (mini tenders). Each phase has its own tasks and evaluation. All solutions of mini tenders are available for all suppliers so they can use, implement and improve them. The goal is to identify the best solution based on work of all suppliers.

Challenges, risks and solutions to overcome obstacles

The successful introduction of PCP in the Czech Republic requires the implementation of several activities. The first step is to create an Action Plan for PCP and PPI. Its aim is twofold: to encourage industry to deliver innovative goods and services on the one hand, and to supply public bodies and citizens with advanced and efficient goods and services on the other. The responsibility for the ongoing implementation will be on the Ministry of Industry and Trade or the TACR.

The biggest obstacle to spreading PCP is the legal framework. According to the legislation prepared (draft of the new act on public procurement and prepared novelisation of the act on Research and Development Support from Public Funds) the PCP concept shall be defined and implemented into Czech legislation. In addition to that, the procurers can have a use of the methodological framework developed by TACR (see above).

The next problem is the lack of information among potential procurers and suppliers. It would be useful to build a service point for PCP (probably as part of the TACR or with the Ministry of Industry and Trade). This service point could act as a contact and information point for PCP questions and would be able to offer services in the field of further education and training, events, pilot projects, monitoring and documentation and PCP online platform. It would be appropriate to create a so-called brokerage initiative designed to create a systematic exchange of information between the public authorities and companies owned by the state on one hand, and innovative companies on the other hand. This role could be taken by the service point as well.

Another issue that needs to be addressed is fragmentation of demand for innovative solutions of governmental and other public bodies and the lack of information on how to optimally balance the risks and benefits of the contract for the contracting and project implementation. It is therefore necessary to disseminate information on PCP and motivate companies to participate in these schemes in the future.

TACR is trying to overcome obstacles in the legal framework in the practical application of regulations for public procurement. Recently, there were two decisions of the Office for the protection of competition supporting TACR process; however, many obstacles in the legal framework persist even though the RDI policy demands an improvement of the legal framework. It will take some time to evaluate benefits coming from other novelties described above.

Key lessons learned

The main lesson is that it is possible to successfully use PCP concept in practice although it is not incorporated into Czech law.

Denmark*

Strategic framework, action plan and scope for procurement for innovation policy

The framework for procurement for innovation is part of a national procurement strategy. In October 2013 the Danish government launched a "Strategy for Intelligent Public Procurement". The strategy included a number of actions that were intended to strengthen focus on:

- ensuring that public procurement supports public sector effectiveness (e.g. through low prices, total costs of ownership and transaction costs)
- using public procurement to develop innovative and high quality solutions (e.g. through support of immature markets, more clear and flexible public procurement processes and increased use of public-private innovation partnerships)
- supporting sustainability and green growth through public procurement (e.g. through increased use of environmental targets and social clauses).

Procurement for innovation policy in Denmark has a broad scope and covers a variety of innovation-friendly procurement tools (e.g. pre-commercial procurement, functional requirements and public-private innovation partnerships). A formalised innovation partnership is called OPI (*offentlig-privat innovationspartnerskab*) and is synonymous to PPI.

Implementation

The government has decided to implement the revised EU directives on public procurement in a Danish law. The purpose is to enhance clarity and flexibility in the public procurement process, including the use of new public procurement tools.

In 2013 the government established a council (Rådet for Offentlig-Privat Samarbejdewith) with the overall target to promote effective and innovative public procurement (mandate expired in 2015). In 2013, the council published an analysis (*Innovationsfremmende indkøb*) that provided an overview of PPI in Denmark and gave examples of PPI in the United States, Great Britain, the Netherlands and Finland. Related to the analysis, the council also published a PPI guidance targeted to public procurement units.

The "Strategy for Intelligent Public Procurement" includes 29 concrete actions related to public procurement and a number of them are related to procurement for innovation:

- offer financial support to pre-commercial procurement initiatives (through *Markedsmodningsfonden* and concrete projects related to the government's 2012 innovation strategy)

* OECD Survey Part I submitted by Griet Storr-Hansen, Senior Adviser, Danish Agency for Science, Technology and Innovation.

- provide guidance on how to use functional requirements in public procurement (published 2014)
- collect and disperse examples of innovation-friendly public procurement.

Denmark's "Strategy for Intelligent Public Procurement" builds on seven guiding principles for public procurement, and the country considers this strategy one of their best practices. The following principles are included:

1. Make sure that public procurement units have sufficient competences and strategic focus to gain from centralisation, synergies and economies of scale.
2. Use dialogue to gain knowledge of the market and user needs and be sure to support market competition in both the short and long run.
3. Choose the tendering form that is best suited to the specific situation and that reduce total costs of ownership.
4. Always consider using functional requirements to support innovation and development of more efficient solutions.
5. Always consider using total costs of ownership to use more efficient resources.
6. Support green growth by using energy and environmental requirements.
7. Prioritise implementing and monitoring contracts to make sure that potential gains of intelligent public procurement are realised.

The Danish government published its first national innovation strategy in 2012. The strategy contains 27 policy initiatives on how to increase the effects of public research and development and innovation (R&D&I) investments on growth and job creation. A number of the policy initiatives concern public procurement of innovation (as defined above):

- Initiative No. 2: Restructure the Business Innovation Fund into a Market Development Fund. By using more tenders with functional requirements or pre-commercial procurement the Market Maturation Fund could, as an example, support the public sector to encourage the development of innovative business solutions via its procurement processes.
- Initiative No. 4: Establishing "INNO+", a solid, professional basis for the prioritisation of innovation policy.
- Initiative No. 5: Establish a model for societal innovation partnerships.
- Initiative No. 6: Initiate pilot innovation partnerships in 2013.

Initiatives 4, 5, and 6 concern the preparation and implementation of a new model for OPI based on societal challenges.

Challenges, risks and solutions to overcome obstacles

The 2013 publication from the council (Rådet for Offentlig-Privat Samarbejde) identified five barriers to innovative public procurement:

1. public procurement legislation
2. opposing interests in the public and private sector and internally in the public sector

3. lack of knowledge sharing
4. unclear responsibility for co-ordinating initiatives at the national level
5. risk aversion.

Solutions to overcome the obstacles: A new law on public procurement will enter into force on 1 January 2016 and introduces the new procedure "innovation partnership". The Danish law on public procurement is based on Directive 2014/24/EU on public procurement. The innovation partnership aims at the development of an innovative product, service or works and the subsequent purchase of the resulting supplies, services or works. The expectation is that the innovation partnership procedure increases the acquisition of new innovative goods, services and works. The new procedure is also expected to create a further incentive in the private sector to develop new innovative goods, which can be used by public authorities to improve the general welfare for its citizens.

Measurement and impact assessment

If not already complete, the initiatives launched in the Strategy for Intelligent Public Procurement are continued and monitored by the accountable ministries and agencies. No central evaluation or assessment is planned in a short term perspective.

Estonia*

Strategic framework, action plan and scope for procurement for innovation policy

Estonia has not (yet) developed and agreed upon a strategic framework for procurement for innovation. Elements of demand-side innovation policy are included in some initiatives and programmes, but this is not the outcome of systemic policy implementation. Innovation policy as such is governed by two ministries: the Ministry of Economic Affairs and Communications (MEAC) and the Ministry of Education and Research. Both ministries have developed strategies where the concept of and the need for further elaboration of procurement for innovation is presented: "Knowledge based Estonia 2014-2020" and "Estonian Entrepreneurship Growth Strategy 2014-2020." The Estonian National Reform Programme "Estonia 2020" also indicates a need to transform public procurement regulations into an engine of development in fields important to the state (e.g. innovation).

Estonia does not have yet a stand-alone procurement for innovation action plan, but is working on it. The intention is to develop several instruments that together constitute a set of co-ordinated activities (collectively called "action plan"), rather than compiling one single document.

MEAC and Enterprise Estonia have developed practical guidance material to procurers on how to procure innovation. The guide is complementing the procurement law. MEAC has set up a horizontal taskforce encompassing different ministries. This taskforce fosters the debates about procurement for innovation nationwide.

Enterprise Estonia has designed different support measures to foster the uptake of procurement for innovation. Pilot support measures were launched in the beginning of 2016. Enterprise Estonia and has started awareness raising activities and trainings and has moved on to financial support to concrete procurement for innovations.

Innovation-oriented procurement policies (POPP) are in initial development phase and, mostly take the form of one-off initiatives or specific elements of regular procurement procedures. Estonia is using the definitions of PCP and PPI.

Implementation

Estonia has prepared specific actions supporting procurement for innovation. As mentioned above, Estonia has developed practical guidance and has created a taskforce in 2016. Several activities begin in 2016: training courses, awareness raising events. Although procurement for innovation is strategically yet underdeveloped, there are some practices to report, for instance:

- ELMO project (2011): Estonian Electro mobility Programme. Procurement for innovation for creating a charging network for electric cars

---

* OECD Survey Part I submitted by Sigrid Rajalo, Executive Officer of Innovation Division, Ministry of Economic Affairs and Communications.

- Iris e-project (2001-10): Successful pre-commercial procurement by the Estonian Defence Forces for developing a device that disrupts radio waves and blocks the remote controlled explosive devices starting signals
- Smart Port (2013-16): A new traffic flow management solution organising pre-check in, check-in and line management for ports with multiple ferry operators by providing a holistic and easy–to-understand service for people with trucks and cars.

Challenges, risks and solutions to overcome obstacles

Estonia faced the following challenges in supporting procurement for innovation:

- In developing strategic and horizontal policies for procurement for innovation, one of the occurring significant challenges is the overall lack of horizontal policy implementation (the co-operation between ministries and agencies could always be better).
- There is a lack of long term vision, and too much focus on short-term resources, too much risk aversion, as well as too little awareness and motivation for innovation. There are exceptions; some sectors are performing very well (e.g. ICT). Openness to innovation in the public sector is at very different levels in different fields. The overall openness to innovation especially with regard to procurement practices could be much better.
- The overall procurement practice is rather conservative and risk averse. The greatest challenge is to change existing practices.
- It has been difficult to use structural funds to develop procurement for innovation measures.

As the process to support procurement for innovation is in an early stage, concrete results to show that these challenges have been overcome have not materialised yet. However, Estonia has conducted some exemplary procurement for innovation projects; more projects are planned for the pilot phase. The obstacles mentioned above need to be addressed systematically. Just applying one measure is not enough. Procurement for innovation encompasses great potential, but at the same time requires smart policy development. Therefore, Estonia has come to realise that all the different affecting factors have to be addressed: the legal framework has to enable procurement for innovation (not just in rhetoric, but in practice), procurement practice needs changes, the risks for procurers have to be managed, the same for the bidders, skills to manage the process of procuring innovation need to be developed, appropriate know-how has to be made available to procurers and bidders, higher level (director generals) administrators have to realise the potential and possibilities of procurement for innovation, impact measures have to be developed together with a monitoring system, etc.

Key lessons learned

Estonia submitted the following lessons learned during their early procurement for innovation efforts:

- There is a need for a systematic approach.
- Horizontal policy co-ordination is essential.

- Awareness and acceptance of this policy by policy designers at a higher level is important.
- Good practice examples to illustrate (evidence to support arguments) are needed.

Measurement and impact assessment

There are several discussions on the subject - quantified target for procurement for innovation in Estonia, but there is no agreed hard target yet. MEAC is conducting a study to determine the percentage of innovative procurements in year 2015 and the results are considered as a base line of innovative procurements in Estonia. MEAC is also starting to monitor the innovative procurements in e-procurements system late 2016. A feasibility study was conducted for the design and implementation of demand-side innovation policy instruments.

Estonia will start to monitor innovative procurements late 2016 and criteria for evaluation were proposed by the authors of the study "Feasibility study for the design and implementation of demand-side innovation policy instruments in Estonia."

Finland[*]

Strategic framework, stand-alone action plan and scope for procurement for innovation policy

There is no stand-alone procurement for innovation action plan in Finland; however, the country has an overall national strategic framework with objectives. The implementation takes place through various sectors and sector strategies. This allows ownership and taking into account sector specific characteristics and demands.

In 2008, the promotion of public procurement of innovation and pre-commercial procurement was recognised as an innovation policy objective in the national innovation strategy. In 2010, Finland issued the Action Plan for Demand and User-Driven Innovation which further elaborated the strategic framework for procurement for innovation. Since then, procurement for innovation tool has been included in several policy documents (mostly for specific sectors such as health, ICT, cities, energy, etc.).

The government programme 2015-19 includes for the first time a numerical target 5%, for innovative public procurement. This target is a strong encouragement to conduct procurement for innovation. The scope for procurement for innovation policy is wide; it encompasses both PPI and PCP. The focus is on PPI since it is seen to offer wider possibilities. There are no specific definitions, but the implementation of the EU Procurement Directives is underway and it will clarify the procurement processes for innovation.

A dedicated helpdesk gives advice and consultancy to public procurers around sustainable and cleantech procurements. The helpdesk gives advice to public procurers through all stages of the procurement process, offers tools and guidelines, suggests criteria to be used in procurement, and collects best practices.

Tekes Smart Procurement Programme is a horizontal programme covering strategic areas of Tekes with most promising future business opportunities encouraging PPI and PCP in various sectors and provides earmarked fundingcing for public procurement units. The project runs from 2013 to 2016. In 2009, Tekes launched a Financial Instrument for Public Procurement, which provides funding for innovative solutions.

The most important goal is of the Tekes programme and funding to create opportunities for SMEs; secondly, the goal is to create innovative solutions to public sector challenges. Tekes funds the planning of public contracts, covering 50% of total project costs. In addition to funding, Tekes fosters networking and co-operation. Tekes has also initiated national training for innovative public procurement for 15 largest cities in 2015 and all 20 health districts in 2016.

Implementation

Two examples of projects conducted under the umbrella of Finland's national framework are

* OECD Survey Part I submitted by Kirsti Vilén, Ministerial Adviser, Ministry of Employment and the Economy.

1. The Ely Centre for Southwest Finland - Flood risk warning system

This project aimed at improving a flood-risk management system, including the prediction of floods, damage estimates, and decision making support. The project was structured as a product pilot and seen, with the aim of promoting innovations. The supplier, in close co-operation with the buyer, developed a new solution. The supplier was aware of the pilot-nature, and that the provided results would not need to be fully ready for scaling up. The company carried the responsibility for the product development work, but the buyer was actively involved, commenting and supporting the development work, as a result of which a pilot solution was created.

For the buyer, this process created a new solution for a significant need. A comparable result would not have been achieved in a traditional way. The supplier company appreciated the opportunity to develop a product interactively with the buyer. This considerably strengthened the understanding of customer requirements and improved the end result. In addition, the company benefitted from new contacts, which help when attempting to enter international markets.

2. The Järvenpään Mestariasunnot oy Zero Energy Building

Järvenpään's zero energy project *Jampankivi* nursing facility consists of two renovated houses and a new building. The new nursing home built in 2011 was constructed in accordance with the building regulations coming into effect in 2020. On an annual level, the renewable energy that is produced by the zero energy building and can be transmitted to energy networks equals the amount of non-renewable energy that the building consumes.

During the design stage of the project, efforts were directed towards co-operation with the researchers in the field, technology companies, material suppliers, designers and subcontractors in order to come up with the best comprehensive solutions for life cycle costs. Different groups (users, buyers, supplier companies) benefited. The project has provided references for the suppliers of product parts targeting multi-storey zero energy buildings. The building costs of a zero energy building are about 15% more than those of a normal building, but it is estimated that this extra cost can be recovered in about ten years' time thanks to inexpensive maintenance costs.

Challenges, risks and solutions to overcome obstacles

The following main challenges were identified:

- Time and resources are needed to raise awareness, given that PPI/PCP is new to the public sector. The learning process required for change is a long process.
- New types of skills, working methods and attitude in general in the public sector are required.
- Procurement for innovation means higher risk (financial, technological, political and societal) and there is lack of skills and tools to manage that risk.

Policy and management level support are often key issues when introducing new procurement practices. Financial support plays a role but even more important is learning of new skills. It takes time to adapt to new practices. This is underlined in Finland where public procurement system is quite fragment (e.g. more than 300 municipalities are responsible for the majority of public procurement volume, including health and social services.).

Key lessons learned

Change takes time and requires both bottom-up and top-down measures.

Measurement and impact assessment

The Finnish government has launched a study to tackle the measurement issues around innovative public procurement. This will help to monitor the use of procurement for innovation and encourage more public procures to use it in the development of public services.

France*

Strategic framework, action plan and scope for procurement for innovation policy

The framework of procurement for innovation in France is part of the innovation strategy as a demand-side support tool. The main objective is to support the growth of innovative SMEs by funding the development of their innovations, providing them with access to new markets and quality references. Public procurement is considered a way to increase the public funding of innovation. The innovation policy highlights how public procurement can act as a level for these policy goals.

The development of procurement for innovation has been established as a priority in the 2012 "National Pact for Growth, Competitiveness and Employment". The pact includes a target to award 2% of public procurement volume (from government, public agencies and hospitals, but excluding defence procurement) to innovative SMEs by 2020. This target approximately represents up to EUR 1.4 billion in additional public funding of innovation. The target has not been achieved yet. Innovation has also been added as a new performance indicator for public procurers.

In this context, the French government uses the following definition of procurement for innovation:

- purchases of products not yet brought to the market, especially when the buyer helped finalise the product specifications in order to attend to an unmet need or bring a new and improved response to an existing need
- R&D procurement designed to foster the emergence of solutions
- purchases of products brought to the market during the previous two years, under the condition that they attend to an unmet need or bring a new response to an existing need.

Concretely, the procurement for innovation policy takes the following form:

- requirement for public institutions to insert a section on innovation in their procurement strategies
- awareness-raising on innovation issues with public procurers
- simplification of the relationship between public purchasers and innovative companies.

Implementation

Since 2012, several actions have been identified and implemented to support public procurement of innovation. These measures were primarily aiming at informing public procurement organisations about innovation, and facilitating the encounter between offer (innovative SMEs) and demand (public institutions). Different ministries are responsible for these actions:

* OECD Survey Part I submitted by David Adolphe, DGE, Administration – Ministry.

- annual roadmap for procurement for innovation, for each ministry and public institution
- "Public procurement innovation" guide, distributed to all public procurers
- online platform to put SMEs in contact with public buyers
- network of procurement for innovation officers in each region to increase the awareness of public stakeholders and SMEs on procurement for innovation issues
- events where innovative SMEs presented their innovative products to public procurers
- training programme on procurement for innovation dedicated to public procurers
- creation of an innovation unit within UGAP (Union des groupements d'achats publics), the central procurement structure for public procurement.

France considered the following among its most successful practices:

- **Creation of an Internet platform**, "Procurement of innovation": This platform was created by the government to facilitate relationships between ministries, public institutions and innovative SMEs. It enables companies to present their innovative products, to achieve direct contact with public procurers and to clarify the needs of public procurers before drafting the product specifications.
- **Promotional events** on procurement for innovation supported matching procurers' needs with innovative solutions. This practice is especially interesting with regard to two key aspects: 1) providing innovative companies the opportunity to meet government buyers and discuss their needs for innovative solutions; 2) enabling public buyers to identify new potential suppliers for a public market and clarify the definition of specifications after the meeting.
- **Creation an innovation unit** within UGAP. This new strategy (2014) will have an important impact on the growth of procurement for innovation. This structure has a significant ability to sign contracts for procurement for innovation and has a dedicated team working on this topic. Procurement for innovation through UGAP reduces tendering time and associated risk.

Challenges, risks and solutions to overcome obstacles

France encountered the following obstacles in supporting procurement for innovation:

- Raising awareness of public procurers about innovation: Buyers should be able to identify innovative products or services and extend their knowledge on existing or future innovations.
- Reducing the risk aversion of public purchasers on innovative solutions, including the legal risk concerning the qualification of "innovative product".
- The development of specific training on procurement for innovation as well as future events promoting procurement for innovation will help tackling these challenges.
- Another important challenge will be the development of shared monitoring systems among the buyers so as to measure procurement for innovation.

Key lessons learned

The measures which have been implemented in France's strategic framework are still too recent to assess. However, one of the main lessons learned is that this policy to support to innovation needs a strong political will to enable procurers to include innovation as a procurement strategy.

Measurement and impact assessment

The Department of State Procurement (*Service des Achats de l'Etat* - SAE) ensures that public procurements are efficient from an economic standpoint, respect the objectives of sustainable development and social development, and contribute to supporting innovation. Since 2012, the government has given priority to the development of procurement for innovation with a target of 2% of the volume of public procurement awarded to innovative SMEs by 2020. Since 2014, the SAE has included in its procurement performance measurement system an innovation indicator. The indicator relies on two ratios:

- Total amount of procurement for innovation contracts awarded by the department / Total procurement from the Department excluding defence and security.
- Total amount of procurement for innovation contracts awarded by the Department for SMEs / Total procurement of the Department excluding defence and security.

The indicator is reported by the buyer at the time of purchase. For the time being there are no impact assessments, evaluation studies and/or studies of state of play regarding any type of procurement for innovation.

Germany*

Strategic framework, action plan and scope for procurement for innovation policy

Procurement for innovation is part of the overall innovation strategy of the German federal government. The "High-tech-Strategy Germany" encompasses all research, technology and innovation measures of the German government. Innovative procurement is the most important measure under the framework of demand oriented policy instruments. The overall strategic goal is to encourage public procurers to buy more innovative and sustainable products. Recently, the ministers for economic affairs of the federal states (*Länder*) also decided to put stronger emphasis on innovative and sustainable products and services in public procurement.

Germany applies a broad definition of innovation including non-technological products and services. Germany distinguishes innovative procurement processes such as electronic procedures and innovative products (PPI). Some German PCP projects have been started.

Implementation

Since 2014, a competence centre for innovative procurement (KOINNO) is established to fulfil several tasks:

- giving advice to procurement offices
- collecting good examples
- building up a database for innovative products, services, procedures and other solutions, which could be used by procurers, as well as information to areas where innovative solutions are required
- conferences for exchange of good practice, including an annual conference of the Federal Ministry for Economic Affairs and Energy (BMWi)
- award by the BMWi for the procurement of an innovative good/solution and for an innovative procurement process
- support of a pre-commercial procurement case as a model.

As a consequence of the 2016 evaluation of the competence centre for innovative procurement (KOINNO) it is intended to increase the work on public relation, individual case consultation and the involvement of the professional decision making level (mayors, head of departments, ministers). KOINNO will be continued in 2017.

Germany's federal government recently published a guideline for procurers that helps them use instruments to foster innovative procurement.

* OECD Survey Part I submitted by Wolfgang Crasemann, Head of Unit, Federal Ministry for Economic Affairs and Energy (BMWi).

Challenges, risks and solutions to overcome obstacles

Germany faced the following challenges in supporting procurement for innovation:

- Lack of political support.
- Lack of qualified staff, who think more strategically instead of regulation-oriented.
- The mentality and risk aversion: Public procurers mostly prefer well-known products and services, not new ones; they hesitate to exchange with businesses and to learn about new technologies. The only way to overcome this is to continuously provide information and training.
- Financial restrictions: Mostly, the budget for investment costs is separate from the budget of operating costs. Consequently, the advantage of an innovative product is not always obvious for the procurer.
- Complexity of the instrument of PCP and the costly measure to start a competition between companies and research organisations: Procurers have to finance more than one organisation. For them, it is not obvious that these higher costs at the beginning will pay off in the future.

While the competence centre works towards overcoming these obstacles, challenges remain.

Key lessons learned

Germany's main lessons learned are that: it is important to continue with the ongoing measures; and to increase activities if more funds are available (high multiplier effect of the budget.)

Measurement and impact assessment

Germany uses evaluation studies.

It is extremely difficult to set quantitative targets. There is no survey about the amount of innovative procurements in Germany. However, a study to investigate possibilities to survey the most relevant statistical data was launched.

Greece*

Strategic framework, action plan and scope for procurement for innovation policy

Greece does not yet have an procurement for innovation action plan. However, Greece is taking first measures addressing this issue. Greece's smart specialisation strategy (RIS 3) 2014-20 includes a programme on Pre-commercial Procurement, conducted by the General Secretariat for Research and Technology (GSRT) and the Ministry of Education, Research and Religious affairs. The programme has a budget of EUR 40 million. A pilot is under preparation.

Greek law defines PCP as follows: "Pre commercial procurement: Buying research services in case the contracting authority or entity does not assume all risks, the results and use benefits in the conduct of its activities, but shares them with the providers under market conditions. The object of the contract falls within one or more categories of research and development defined in the present context. The contract is of limited duration. With the exception of a prototype or a limited set of first test / validation data, the purchase of goods or services, which are developed within the framework of a pre commercial procurement, should not be the subject of the same contract."

(It is noted that this definition maybe amended during the preparation of the first PCP pilot taking into account best practices of other countries as well as the results of the incoming consultation.)

Implementation

There are no specific actions to support procurement for innovation at the national/central/federal level.

Challenges, risks and solutions to overcome obstacles

Greece identified the following challenges, which have yet to be overcome:

- achieving co-ordination among ministries
- overcoming administrative and training obstacles
- identifying appropriate personnel who will undertake the execution of Public Innovation Strategy at different levels.

Measurement and impact assessment

There is no system in place to measure the impact of actions related to procurement for innovation and there are no quantified targets for procurement for innovation in Greece. Impact assessments, evaluation studies and/or studies of state of play regarding procurement for innovation do exist.

* OECD Survey Part I submitted by Konstantinos Tzanetopoulos, Head of Development and Coordination Department, General Secretariat of Commerce and Consumer Protection.

Hungary[*]

Strategic framework, action plan and scope for procurement for innovation policy

Public procurement promoting innovation has a high position on the Hungarian policy agenda. Both on the European and the national level, it is regarded as an important demand-side instrument. While there is no stand-alone procurement for innovation action plan yet, procurement for innovation is addressed in Hungary's part of the EU strategy EU2020 as well as in the "Investment into the Future" - The National Research and Development and Innovation Strategy 2013-20 and in the National Smart Specialization Strategy (S3). The S3 has Pre-commercial Procurement as one of its pilot projects to be carried out in 2017-18.

In correlation with the new European directives, the revised Hungarian Act CXLIII of 2015 on Public Procurement has two new procedures particularly relevant for authorities that wish to purchase innovative goods, services or works: the innovation partnership and the competitive procedure with negotiation.

In accordance with the European directive and guidance on public procurement of innovation, the main idea behind any policy-based effort in Hungary is to allow greater scope for interaction and dialogue with the market once the public authority's need is articulated in a particular category or to solve a specific challenge.

Boosting innovation through demand-side measures is not a novel approach, reflected by the fact that policies such as technology-based standards or innovation-oriented regulations have been around in several sectors. However, the focus has shifted to the use of procurement for innovation because this issue already has a key role among the measures supporting the implementation of EU objectives. By applying procurement for innovation, savings can be achieved in the procurement budget spent on an annual basis, and the money saved can be allocated to programmes that fund RDI; the quality of public services can be improved through demand-driven, tailored procurement structures; micro, small and medium-sized enterprises will be assisted in reaching new tenders, which have been proven to be unattainable for them in the past. Hungarian policymakers tend to stick to and use the European Commission's definitions, such as PCP, PPI, Forward Commitment Procurement and Small Business Research Innovative.

Implementation

Hungary submitted the following successful cases:

- The project RAPIDE - bringing innovative products and services to the market more quickly - was an example of a RFEC project active on pre-commercial procurement. The RAPIDE Innovative Procurement Working Group led by *Észak-Alföld* Region (HU) piloted the application of structural funds resources to set up pre-commercial procurement programmes.

[*] OECD Survey Part I submitted by Dávid Lakatos, Office referent, National Development, Research and Innovation.

- The goal of PROGR-EAST was to encourage the use of pre-commercial procurement (PCP), in five targeted European countries (Poland, Czech Republic, Slovak Republic, Hungary and Slovenia), taking into consideration the specific needs and constraints for public procurement, and in particular for the procurement of R&D existing in Eastern Europe. The aim is to introduce innovative PCP strategies to public authorities, universities and industrial stakeholders, and transfer successful experience implemented in other European and external regions to implement innovative public services. The Hungarian participant was the *Puskas Tivadar* Foundation CERT-Hungary.

The INNOVA *Észak-Alföld* Innovation Agency is currently taking part in two projects: The Smart@Fire project and the iMAILE project.

In addition the *Észak-Alföld* Regional Development Agency (AGENCY) has been involved in the P4ITS (http://p4its.eu/), a thematic network gathering contracting authorities experienced or planning to shortly embark on deploying Cooperative Intelligent Transport Systems and Services (C-ITS), and willing to improve the market roll-out of innovative transport systems and services through PPI. However, the agency was dismissed by 31 March, 2016, and was succeeded by the Department of Development, Planning and Strategy of the *Hajdú-Bihar* County Council (department). Furthermore, it was the department, which participated in the finalisation of the Final Recommendations/Guidelines of the project.

Challenges, risks and solutions to overcome obstacles

The public organisations in Hungary lack the necessary financial and human resources to initiate procurement for innovation procedures like PCP or PPI. H2020 projects do have potentials for Hungarian participation and a small number of public actors are interested in joining cross-border PCP/PPI projects, but mainly as observers. Hungary has not yet been able to overcome the above-mentioned obstacles.

Key lessons learned

For Hungary, a key lesson learned is how public organisations tend to stick to old routines and are very distrustful of new procurement processes. This is particularly true of procedures such as PCP or PPI that are not yet regulated by any act or government decision.

Measurement and impact assessment

Hungary does not yet have a system in place to measure the impact of actions related to procurement for innovation. The system will be developed in the course of the PCP pilot programme under the National Smart Specialization Strategy.

Ireland*

Strategic framework, action plan and scope for procurement for innovation policy

As part of the Public Service Reform Plan, the government is reforming the public procurement process to deliver greater value for money through increased use of common procurement frameworks, centralised purchasing, increased professionalism and more innovative use of technology. The government has established an Office of Government Procurement (OGP) as an independent body under the aegis of the Department of Public Expenditure and Reform to drive a new consolidated and integrated approach to public procurement.

There is no stand-alone action plan. Innovation is one part of Ireland's overall procurement strategy. The intention is that the OGP works with other public sector departments and support opportunities that encourage innovative procurement.

The scope of procurement for innovation is considered by Ireland's Category Councils. The OGP has established Category Councils for 16 categories of goods and services bought by the Public Service. Category Councils are responsible for developing commercial strategies for sourcing goods and services in each of their categories in line with the needs of customer organisations and in the context of obtaining best value for money.

Implementation

Ireland's specific activities related to procurement for innovation are based on Ireland's Programme for Government (2011-16), which recognises the role that public procurement can play in supporting innovation. It commits to reform public procurement to become a tool to support innovative Irish firms and to allow greater access to Irish small and medium sized businesses.

The government's Action Plan for Jobs, which is the responsibility of the Department of Jobs, Enterprise and Innovation, also recognises that procurement can be an enabler of private sector innovation and can support the drive to reduce costs in procurement budgets. It acknowledges that many innovative companies can offer solutions to the needs of public sector bodies with lower whole of life costs than more conventional purchases. It recognises the need for contracting authorities in Ireland to be become more open to procuring innovation. It commits to examining practical ways to highlight the merits of purchasing innovative products and services, where appropriate, as a means of achieving cost savings in public procurement.

Responsibility for green procurement rests with the Department of Environment, Community and Local Government. Green Tenders, An Action Plan on Green Public Procurement (GPP) recognises that GPP can be a driver for innovation and competitiveness in the industrial sector promoting the development of new technologies and providing a competitive advantage for emerging companies in the SME sector. Public

* OECD Survey Part I submitted by Karl Ryan, Assistant Principal Officer, Office of Government Procurement.

procurers are asked to keep themselves informed about changing technologies and processes, so that where these are cost-effective and fit for purpose, relevant developments and innovations can be included in the procurement process. This is reinforced in the policy paper Green Procurement Guidance for the Public Sector which suggests that consideration of GPP in the pre-procurement phase should include amongst other things re-thinking how the demand can be met in a way that requires fewer or better value goods to be bought and is an ideal time to identify and evaluate new innovations in the market place.

Public procurement in Ireland is underpinned by a number of core principles, in particular the need to maximise competition in the market for the goods and services purchased by the State. Where practical and legally possible, policy also seeks to promote whole-of-government objectives, including the promotion of innovation in procurement. It is the stated aim of the OGP not just to provide value for money but to provide sourcing solutions that are smarter and more efficient.

The OGP has an SME Working Group with representatives from industry representative bodies and the Department of Jobs, Enterprise and Ireland. The focus of this group is to develop and monitor strategies for SME access to public procurement. Support for innovation falls broadly within the terms of reference of this Group.

The OGP conducts a targeted programme of seminars, workshops and "Meet the Buyer" event which afford suppliers an opportunity to meet and discuss the issues with public service buyers including innovation issues and/or proposals.

One of Ireland's best practices relates to guidance. Procurement for innovation policy is addressed in DPER Circular 10/14, titled "Initiatives to assist SMEs in Public Procurement," where it is recognised that innovation and creative solutions can assist procurement strategy. A comprehensive approach to procurement for innovation is provided in "Buying Innovation - a 10 Step Guide," which was published by the Procurement Innovation Group in 2009. It highlights the role that SMEs can play in smart and innovative procurement.

Under the government's Action Plan for Jobs for 2014, the government committed to introducing, on a pilot basis, a Small Business Innovation Research (SBIR) programme to provide opportunities for innovative solutions to be developed to meet the needs of public bodies. In this regard on 30 June 2014 the Sustainable Energy Authority of Ireland (SEAI) in co-operation with Enterprise Ireland and ESB launched Ireland's first SBIR competition. SBIR falls under the category of pre-commercial procurement (PCP). The competition's first target is smart solutions for charging electric vehicles (EVs) in communal parking areas such as apartment blocks. Up to EUR 200 000 will be made available to develop the prototype. The proposed solution should be adaptable to cater for underground, multi-storey or other privately run car park developments where EV owners will require access to charging facilities.

Challenges, risks and solutions to overcome obstacles

At ground level it is the responsibility of the newly-established Category Councils to examine the role that innovation can play in their procurement strategies and to examine where it is appropriate and relevant. The OGP is a new organisation. It is too early to elaborate on innovation challenges at this stage.

There is a clear focus within the OGP to foster and facilitate innovation in public procurement; this is expected to help overcome obstacles for procurement for innovation.

Key lessons learned: It is too early to elaborate on lessons learned at this stage.

Measurement and impact assessment

The reason why no system currently exists to evaluate the impact of procurement for innovation is that the reform of public procurement is in its early days and procurement innovation is presently just being tested on a limited basis. It would be the intention to evaluate procurement innovation at a later stage.

Italy*

Strategic framework, action plan and scope for procurement for innovation policy

At the time of the OECD Survey, the Ministry of Education, University and Research (MIUR) was about to pass the National Research Plan (NRP) 2015-20. This plan aims at streamlining, simplifying and boosting the national research system by sustaining all research phases with new funds and by offering a stable and innovative policy framework. The NRP fosters policies that sustain research through the promotion of public demand for innovative solutions, and makes pre-commercial procurement an integral part of Italian national research policy.

There is no stand-alone procurement for innovation action plan. At the moment, the most complete normative and organisational framework is defined by the guidelines issued by the Dipartimento per la digitalizzazione della pubblica amministrazione e l'innovazione tecnologica and the Ministry of Education, University and Research (MIUR) in 2012. Following these guidelines, the Ministry of Education, University and Research (MIUR) and the Agency for Digital Italy (AGID) recently defined a pre-commercial procurement programme for the procurement of research, development and innovation services by public bodies. Even though this programme is of considerable scope, it cannot be considered a comprehensive national procurement strategy.

The pre-commercial procurement programme promoted by the Ministry of Education in co-operation with AGID has a total value of EUR 100 million. The calls for proposals include topics such as new solutions and cloud services for the automation of administrative procedures and document management; multi-parametric monitoring systems and the correlation of multimodal events for the preventive alerting of natural disasters and rapid organisation of emergency interventions. All actions currently undertaken or designed by MIUR refer exclusively to pre-commercial procurement, as defined by the European Commission.

Implementation

The MIUR-AGID agreement is a recent successful experience aimed at promoting research and development based on the demand for innovation expressed by the Public Administration (PA). Besides this pre-commercial procurement programme, there are valuable local valuable experiments, for example in Lombardy and in Puglia regions: In 2012, Lombardy Region and MIUR, with the support of The European House - *Ambrosetti*, launched a technical dialogue with the market, followed by a pre-commercial procurement initiative in the health sector. The Puglia region issued a call in the "independent living" sector, aimed at finding innovative solutions to improve the quality of independent life for people who are not self-sufficient.

* OECD Survey Part I submitted by Mario Calderini, Senior Advisor to the Minister for Research and Innovation Policies, Ministry of Research and Innovation.

Challenges, risks and solutions to overcome obstacles

The first challenge has to do with the normative framework of public tendering. At the start, the efforts on public procurement have been focused on rationalising it by the centralisation of all purchases in Consip S.p.A.. More recently, since 2012, "judicially validated" guidelines and templates were defined and allow (IPP) calls by Italian administrations.

Secondly, geographic fragmentation is a challenge. Local administrators prefer to engage in research and innovation policies with their resident constituency, while IPP calls need to be open on a wider scale. As IPP policy build on traditional local innovation policies, there is a geographical bias. Besides, the fragmentation of resources across many local actors may weaken the scale effects of innovative procurement.

Third, industrial stakeholders' preference for demand-driven innovation policies is a challenge.

These obstacles have been partially overcome by defining national guidelines in line with the legal framework.

Key lessons learned

Italy submitted two lessons learned. First, innovative policy making benefits from an experimental innovation policy approach, and has to integrate monitoring and impact procedures. Secondly, co-ordination and joint strategic planning is strongly required in order to avoid geographic biases and resource dispersion.

Measurement and impact assessment

There is no system in place to measure the impact of actions related to procurement for innovation in Italy. All actions regarding IPP impact evaluation are referred to the NRP 2015-20. For the planning and implementation of the policies and tools provided by the NRP, within the General Directorate for Research, a unit will be established that is tasked with monitoring. This unit is also tasked with providing access to evidence processed by the different parts of the research system and reprocess them in order to conduct evidence-based modelling. Moreover, this action will develop procedures and tools for ongoing monitoring and analysis of information sources that will allow reproducing technological foresight analysis and documents, acting in synergy with ex-ante and ex-post assessments of research policies.

Korea*

Strategic framework, action plan and scope for procurement for innovation policy

Korea's strategic tools for procurement for innovation include:

1. Set-aside for new technology products
   - set-aside for products manufactured with domestically developed new technology
   - qualifying products are given a certification (different certifications by product categories)
   - central and local governments and public enterprises should fulfil 20% of their procurement with innovative procurement
   - new technology-certified products are expected to account for 20% of the procurement of the specific product type for which new technology products are available, not 20% of total procurement.
2. Set-aside for SME-developed technology products
   - Set-aside for products manufactured with new technology developed by an SME.
   - Qualifying products are given a certification (different certifications by product categories).
   - Central and local governments and public enterprises should fulfil 10% of their SME-product purchase from SME-developed technology products (10% of the procurement of the specific product type for which SME-developed technology products are available, not 10% of the total procurement). Public entities must fulfil 50% of their procurement from SMEs where SME products are available.
3. Designation of "Excellent Government Supply Products" and allowing direct purchasing for them
   - Public Procurement Service (PPS) designates outstanding SME products in seven categories (ICT, electric and electronic, construction and environmental, chemical and fabric, machinery, office equipment, science and metical), through application and evaluation process.
   - Public entities are allowed to purchase Excellent Government Supply Products through direct contracting.
   - The period of designation is three years. As of July 2015, a total of 1 128 product models are designated as Excellent Government Supply Product.

* OECD Survey Part I submitted by Kang-il Seo, Deputy Director, International Cooperation Division, Public Procurement Service.

There are no other specific actions than specific certification programmes such as NEP (New Excellent Product), NET (New Excellent Technology), GS (Good Software), etc.

Implementation

Specific examples of policy instruments used to support procurement for innovation as described above.

Challenges, risks and solutions to overcome obstacles

Matching/customising certification programmes to suit the actual/future public procurement demands.

Key lessons learned

There is the need to strengthen the studies on current and future demands in public procurement to better harmonise the certification programmes and the actual public procurement demands.

Measurement and impact assessment

There is a system in place to measure the impact of actions related to procurement for innovation and its quantified target. For impact assessments, Korea uses evaluation studies.

Lithuania*

Strategic framework, action plan and scope for procurement for innovation policy

There is growing attention towards demand-side innovation policies in Lithuania. Lithuania's procurement for innovation action plan is part of the country's general innovation and procurement strategy. The importance and potential of procurement for innovation is stressed in the Lithuanian Innovation Development Programme 2014-20. The necessity of the model of pre-commercial procurement was stated in the programme. The Ministry of Economy published Guidelines on Innovative Public Procurement. These guidelines describe how public procurers can buy goods, services or works of better quality, more adapted to their needs, services or goods that could enhance performance of public procurers and quality of their services, and increase demand for innovation on the market. To add, the Ministry of Economy has drafted and the government of Lithuania in 2015 has established the description of pre-commercial procurement. The survey of public purchases for pre-commercial procurement has been carried out recently and the need of approximately 80 pre-commercial procurements were recognised.

Lithuania follows EU strategies and definitions.

Implementation

Lithuania focusses on promoting PPI and PCP. A series of seminars is organised to improve public procurers' understanding of PPI and PCP. Individual consultations are being carried out as well. However, guidelines on PPI and PCP are not considered the main instruments for increasing the number of procurement for innovation cases in country. Lithuania plans to develop methodologies related to PPI and PPC, which will provide procurers with all relevant and guiding information. A series of seminars and matchmaking events for public sector institutions and businesses are being organised in the near future. A financing scheme for PPC supports these efforts.

Challenges, risks and solutions to overcome obstacles

Lithuania observed a set of challenges in its procurement for innovation efforts. For example, contracting authorities (CAs) avoid carrying out innovative public procurement because of additional risks related to the acquisition of innovative solutions. CAs also lack competences and experience in this type of procurement. There is insufficient guidance on the implementation of innovative public procurement. While there are no legal obstacles in public procurement framework to use PPI by CAs, the traditional practice of using price as the single criterion for awarding contracts is dominating. It is more legally sound for CAs to follow more prescriptive arguments set in legislation and avoid the uncertainties of innovative actions. Even if CAs have strong arguments in favour of innovative practices, and determination to conduct procurement for innovation, they face pressures to prove that their choice is made without an intention to restrict competition and is not corrupt. Lithuania plans to solve these issues by adopting

* OECD Survey Part I submitted by Aurelija Kazlauskienė, Head of Innovation policy division, Ministry of Economy.

recommendations, and by implementing demand-side policy tools in the 2014-20 financial period. There is high interest and need for innovative solutions. Therefore, while educating public procurers, Lithuania increases competences and understanding of the subject which will lead to the increase of use of PPI and PCP.

Key lessons learned

It is important to note that public procurers are keen on innovative solutions. However, public procurers often lean to the status quo, due to a lack of competency and understanding. Education and training improves capabilities, but also changes public procurers' mind-set towards in a way that they seek out innovative and more effective solutions.

Measurement and impact assessment

The procurement for innovation has just begun to gain in importance. Therefore, up until now, there was no need to set up a system to evaluate the impact of procurement for innovation. With the basic regulation for PCP in place, Lithuania plans to monitor the implementation of this type of procurement and to measure the impact in the country.

In 2011, Lithuania's the Public Procurement Office started collecting statistics on the number of innovative public procurement cases in the country. According to the Innovation Development Programme 2014-20, the share of innovative procurement should have accounted for 2% of all procurement in 2017 and 5% in 2020. A study evaluating the status quo and providing guidance for drafting legislation in the field of pre-commercial procurement was carried out and approved in 2013. The study relies strongly on best practice examples such as in Norway, the United States. and the Netherlands. The study examines in detail the process of pre-commercial procurement, issues related to IPR and the roles of all actors involved in procurement.

Malta*

Strategic framework, action plan and scope for procurement for innovation policy

Malta has limited experience with procurement for innovation, and does not have a stand-alone procurement for innovation policy. However, existing procurement structures do allow for procurement for innovation to take place and a number of examples of the application of procurement for innovation exist.

Procurement for innovation is seen as a means to obtain improved public products and services while stimulating demand for innovation in the local economy. Procurement for innovation is identified as an important tool to stimulate demand-side innovation in Malta's new National Research and Innovation Strategy 2020 and possible actions will be included in the National Research and Innovation (R&I) Action Plan, which is currently being developed. Regular awareness raising sessions with various types of procurers are held to highlight the advantages of procurement for innovation. Changing the present procurement culture is generally identified as a major obstacle.

Implementation

The Malta Council for Science and Technology and the Department of Contracts within the Ministry of Finance are the main bodies in Malta responsible for promoting procurement for innovation. Most of the work focuses on raising awareness of the benefits of procurement for innovation, helping to steer procurers away from the more traditional, fixed procurement procedures. To this end, the Malta Council for Science and Technology, together with the Department of Contracts, has set up a working group to assist such organisations to engage in innovative procurement. The aim is to offer the necessary support to identify the organisation's innovation needs and the process involved in procuring it.

In addition, the Malta Council for Science and Technology in collaboration with the Department of Contracts has embarked on seven competitive dialogue processes simultaneously. These processes intend to push for procurement for innovation and to break from the traditional services, supplies and works tenders awarded via the cheapest technically compliant tenders. The competitive dialogue processes have so far proven to be highly successful and to date two processes have been awarded whilst other five are in their final stages and in the proximity of being awarded. The subject matter of the competitive dialogues are processes from design to fabrication to installation of interactive science exhibits for the National Interactive Science Centre, which include the design to installation of a Planetarium, the first of its kind in Malta.

There are examples of successful stand-alone cases of procurement for innovation in Malta that can be considered good practice. A number of these projects were documented as part of Work Package 4 of the ERA-PRISM FP7 project which looked at the small-county dimension of procurement for innovation. The case studies focused mostly on innovation in the procurement process itself.

* OECD Survey Part I submitted by Jacqueline Barbara, Strategy and Policy Executive, Malta Council for Science and Technology.

Challenges, risks and solutions to overcome obstacles

In Malta, the most significant challenge is the need for changing the mind-set of procuring entities, away from the traditional procurement procedures to procurement for innovation. A wider understanding of the procedures and processes involved coupled with a greater understanding of the benefits of procurement for innovation would also result in greater support at all management levels to the use of such procedures.

While personnel at management levels are becoming more aware of the benefits of procurement for innovation and how it can be utilised, the obstacle is still in place and it is still considered the largest challenge.

Key lessons learned

Malta considered a lesson learned that a wider understanding of the procedures and processes coupled with a greater understanding of the benefits of procurement for innovation would eventually result in greater support at all management levels to the use of such procedures.

Measurement and impact assessment

Since procurement for innovation is still at its infancy, Malta focuses on raising awareness about the use of procurement for innovation and changing the current procurement system. A second step will be to undertake proper evaluations of the impact of procurement for innovation.

Mexico*

Strategic framework, action plan and scope for procurement for innovation policy

In 2013, President Enrique Peña Nieto instructed the Ministry of Economy to create a programme to drive innovation through public procurement. In order to fulfil this mandate, the Ministry of Economy, on co-operation with other relevant entities, will concentrate efforts on:

- promoting innovation, especially within micro, small and medium-sized firms
- improving public services through innovative products and services.

In pursuing these objectives, the Ministry of Economy will adhere to a three-phase plan:

1. Short-term: Design and pilot the innovation on small scale.
2. Medium-Term: Implement the innovation programme in the federal government.
3. Long Term: Promote a culture of innovation and scientific development.

The procurement for innovation action plan is part of the country's general innovation or procurement strategy. In pursuits of its objectives, the Ministry of Economy will adhere to an above-mentioned three-phase plan.

The scope: During the design of the programme the following definitions were established:

- Commercial procurement.
- Regular procurement for innovation: The contract is awarded to the most innovative tender (evaluated by criteria that address innovation in the entire cycle of the good or service).
- Technological procurement for innovation: Public procurement of good or services that are not yet available at the moment of the solicitation but that can be developed within a reasonable time.
- Pre-commercial procurement: Procurement of research and development involving risk-benefits sharing between the government and the tender in order to develop innovative solutions not available in the market.

Implementation

In 2015, The General Directorate of Information Technology of the President's Office, in collaboration with other agencies of the Federal Public Administration, launched the project "Public Challenges". Through the retos.datos.gob.mx platform, Mexican companies were invited to compete by offering innovative digital solutions to problems related to environment, health, education, transportation, food, connectivity and

* OECD Survey Part I submitted by Sara Hernández Munoz, Secretaria de Economia, Directora.

prevention by creating digital applications. In total, 15 Public Challenges were launched and 341 proposals were received. In each challenge, five finalists were selected by a non-governmental committee and received a grant to develop a functional prototype and the best one was awarded a contract to fully develop the selected project.

After the results of the Public Challenges, and in order to strengthen the implementation of procurement for innovation policy, the Working Group on Procurement for innovation was created. The main objective of this working group is to generate policies to mitigate the risk of adopt innovation, to propose modifications to our current legal framework for procurement to facilitate procurement for innovation.

Challenges, risks and solutions to overcome obstacles

The most significant challenges identified are our current institutional and legal framework which lack the flexibility and simplification needed to successfully implement large-scale procurement for innovation; and the reticence of the procurers to participate in the project given its novelty and the procurer's wish to avoid the risks they perceive in such procurement policy.

Key lessons learned

In order to enhance the procurer's willingness to participate in an procurement for innovation programme, it is necessary to design mechanisms of risk management to identify and reduce risk and lessen the potential costs of implementing the programme. Also, for all involved parties, a stronger political commitment is needed.

Measurement and impact assessment

Mexico does not have a system in place to measure the impact of actions related to procurement for innovation since the programme is not yet fully established. Nevertheless, information about the "Public Challenges", our small-scale procurement for innovation projects, is available in the platform *http://retos.datos.gob.mx*.

Netherlands*

Strategic framework, action plan and scope for procurement for innovation policy

The Netherlands have a goal to spend 2.5% on innovation.

There is a stand-alone procurement for innovation action plan: the programme "Innovatiegericht Inkopen", which includes the use of PPI, PCP, SBIR and other instruments that stimulate innovation among contracting authorities prior to the actual procurement. The programme focuses on the stimulation of the dialogue between contracting authorities and businesses prior to the actual procurement, the insight in market challenges and demand articulation. Procurement for innovation aims at providing room for innovation in public procurements and actively challenging businesses to deliver innovation.

Implementation

The Netherlands implemented a number of specific actions. Funding by SBIR is used to compile space data for Dutch contracting authorities; green deals and city deals are used to articulate demand; contracting authorities are supported and the dialogue between businesses and contracting authorities is stimulated. Synergies are actively searched between procurement for innovation and secondary policy goals such as sustainable procurement.

Key lessons learned

Netherland's main challenge to procurement for innovation is the willingness of contracting authorities to take and share risks. Contracting authorities with a more positive mind-set to innovation tend to be the first to embrace the concept of procurement for innovation.

Measurement and impact assessment

The Netherlands monitored the above-described 2.5% target. To assess impact, the Netherlands conducts studies of state of play.

* OECD Survey Part I submitted by Floris den Boer, Senior advisor, PIANOo.

New Zealand*

Strategic framework, action plan and scope for procurement for innovation policy

New Zealand is committed to open, transparent and competitive government procurement that: delivers best value for money, does not discriminate against suppliers (whether domestic or international) and meets agreed international standards.

Government procurement is based on the Five Principles of Government Procurement, the Government Rules of Sourcing, and good practice guidance. Collectively, these provide a broad framework that supports accountability for spending, sound business practice and better results. Applying the Rules is essential to providing open and fair competition that supports innovation and helps create a competitive, productive supply base in New Zealand.

The Procurement Functional Leadership and the NZ Government Procurement Branch of the Ministry of Business Innovation and Employment provides a centre-led (rather than centralised) approach to public procurement which is carried-out by government ministries and departments.

New Zealand has an enabling approach towards public procurement fostering innovation. The policy framework provides a flexible and supportive environment for procurement generating new and improved solutions. Guidance tools and templates are available to departments and ministries (e. g. guides and tools for Competitive Dialogues, Cost Benefit Analysis, and Total Cost of Ownership) to improve procurement. Procurement experts are available to support ministries and departments implementing complex or alternative procurements. Ministries and departments are using public procurement to foster innovation as part of their mainstream/normal procurement processes.

The procurement for innovation action plan is part of the country's general innovation or procurement strategy. Good procurement practices should generally foster new or improved solutions (innovations). The New Zealand procurement strategy is focused on improving procurement skills and practices in ministries and departments, and developing an environment in which businesses can succeed. Public procurement is also part of the New Zealand government's action plan "Business Growth Agenda: Building Innovation". Actions include encouraging innovative public sector ICT procurement models, including accelerator programmes and hackathons.

In New Zealand, government support for R&D and innovation and commercialisation is provided through dedicated organisations, such as Callaghan Innovation, New Zealand Trade and Enterprise, University Commercialisation Office of New Zealand, the New Zealand Health IT cluster, and the Medical Technologies Association of New Zealand. Ministries and departments generally do not have budgets for external research and development, and do not have separate budgets for procurement related R&D. Public procurement fostering new and improved solutions (innovation) is done as part of mainstream good procurement practice.

* OECD Survey Part I submitted by Shayne Gray, Acting General Manager, Government Procurement, Ministry of Business, Innovation and Employment.

New Zealand does not have separate programmes such as SBIR, PCP and PPI for procurement for innovation.

Implementation

There are specific actions to support the implementation of procurement for innovation at the national level of government. The Government Rules of Sourcing provide a supportive and flexible policy framework that encourages good procurement practices which can foster the development of new and improved solutions.

New Zealand's government procurement branch has developed procurement guides, tools and templates to assist with implementation good procurement practices. It has also facilitated procurement professional development (CIPS, courses, online training, mentoring, and a graduate intake programme)

New Zealand's Government Procurement Commercial Pool provides expert assistance on the implementation of large, complex, or alternative procurements undertaken by ministries and departments.

A Procurement Capability Index has been developed and is currently being piloted to enable ministries and agencies to assess their practices and identify areas for improvement.

New Zealand provided the following examples for policy instruments used to support procurement for innovation:

- Policy instruments: Government Rules of Sourcing - for the Public Sector
- Financial instruments: Better Public Services, Seed Fund - for Public Sector
- Programmes: Better Public Services, Result Area 9 - for Public Sector
- Stand-alone-cases: Pilot of Accelerator - for Public Sector
- Other: Commercial Pool experts - for Public Sector.

Challenges, risks and solutions to overcome obstacles

The most significant challenges in developing and implementing policies for procurement for innovation have included: the development of good procurement practices and professional development across government ministries and departments; encouraging ministries and agencies to balance risks and benefits; the limited time and resources to implement good procurement practices, such as early market engagement and supplier relationship management, that can generate new solutions and overcoming "myths" of what can and cannot be done in good public procurement.

New Zealand has been endeavouring to overcome these obstacles by undertaking the following activities:

- Government Rules of Sourcing providing a flexible and supportive environment for good procurement practice
- ministries and departments remaining responsible for implementing good procurement practices

- increased transparency of government procurement through Government Electronic Tenders Service (GETS)
- guides, tools and templates and capability development improving professional standards in government ministries and departments
- commercial pool providing procurement expertise to government ministries and departments
- reviews of ministries and departments procurement practices and development of Procurement Capability Index for ministries and departments to assess their practices.

Key lessons learned

Improvements in public sector practices require a clear and consistent policy framework, and on-going encouragement and support, ultimately linked to ministries and departments' performance accountability mechanisms. Supply markets can sometimes be sceptical about public sector intentions to improve practices.

Measurement and impact assessment

New Zealand has the following systems in place to measure the impacts of actions related to procurement for innovation:

- Reviews of ministries and departments procurement practices.
- Development of the Procurement Capability Index against which ministries and departments can assess their procurement practices.
- Better Public Services, Result Area 9, monitoring includes procurement practices indicators.
- An evaluation of the Accelerator pilot is currently underway.
- Individual mainstream procurement projects have identified innovations and impacts.

A target has not yet been quantified for procurement for innovation. Regarding procurement for innovation, evaluation studies and a survey of government suppliers includes questions on openness to innovation.

Norway*

Strategic framework, action plan and scope for procurement for innovation policy

Procurement for innovation has been a priority of the current government since 2013 and was formulated in the government platform. Implementation is the responsibilities of the ministries and agencies. The previous government had a separate action plan (2013). This action plan is was voted and approved by parliament as a measure in a white paper.

There is no stand-alone procurement for innovation action plan. The agency for Public Management and eGovernance (Difi) and the National Programme for Supplier Development have developed a national method for procurement of innovation. The method gives public purchasers a systematic approach when conducting procurement for innovations. The method encourages extensive dialogue with the market. Procurement for innovation in Norway is to a large extent associated with this method. Several public organisations conduct pre-commercial procurements. EU definitions such as PPI and PCP therefore serve as useful terms of reference although concrete national schemes might have slightly difference delineations of terms.

Implementation

Specific actions: The Ministry of Trade, Industry and the Fisheries is responsible for formulating policy in this field at national level, but a number of ministries, agencies and municipalities have been developing their own initiatives and implementation is distributed. This combination of both central and decentral policy formulation is likely to continue. A number of general schemes originate in the ministries. Specific projects tend to originate in the relevant agencies (first and foremost, agencies responsible for infrastructure in healthcare).

A working group has been established in order clarify roles, co-ordinate actions and to make it easier for public purchasers to find orientation among agencies supporting procurement for innovations. The working group consists of representatives from Innovation Norway, Difi, The Research council and the Supplier development programme. The agency for Public Management and eGovernance (Difi) has a special assignment in order to assist the governments work in renewing the public sector. One of the tasks is to help agencies and authorities to set up solid public procurement procedures. Difi gives guidance and courses in regards to procurement for innovations based on the method developed by Difi/Supplier Development programme, as well as EU methods (i.e. pre-commercial procurements).

There is also co-operation between industry and the public sector about policy development, for instance through the National Programme for Supplier Development. In addition there are co-operation at the Nordic Level, through sector specific programmes in for instance healthcare and building.

* OECD Survey Part I submitted by Eivind Lorentzen, Specialist Director, Norwegian Ministry of Trade, Industry / Department of Research and Innovation.

Best practice:

- The Innovation Norway's Research and Development Programme "Industrial and Public Research and Development Contracts" (IRD/PRD) is a strategic support programme for the industry and public sector. The award-winning financial instrument is set up to stimulate user-driven innovation based on a binding agreement between a public sector entity and innovative Norwegian SMEs. The objective is to support development of new solutions and more innovative procurement. (Annual support to PRD projects approximately EUR 8 million and approximately 60 projects).
- The "National Programme for Supplier Development" is intended to encourage innovation and creativity within public procurement through concrete procurement for innovation projects, method development and competence building activities. The programme has five-year duration period and started in 2010. It will be continued. The Confederation of Norwegian Business, NHO, and Local government interest- and employer organisations KS are the initiators to the programme which is implemented with a partnership of national innovators, state enterprises, greater local councils and the business sector. The programme is supported by the ministries.
- The National Programme for Supplier Development has conducted about 40 pilots (procurements of innovation) based on the national method for procurement for innovation. The pilots have been systematically evaluated and experiences are communicated to Difi in order to further develop the method.

Challenges, risks and solutions to overcome obstacles

The main challenge in Norway is moving from good cases to general practice. There are two aspects to this challenge:

- improving procurement practice in general so that it becomes more innovation enhancing
- spreading specific innovations resulting from innovative procurement processes to the relevant public users, i.e. improving the uptake of resulting innovations.

In order to overcome these obstacles Norway highlights the following areas for change:

- Public procurement must to a larger extent be considered as strategic tool to manage and to improve the public sector.
- General competence on how to conduct procurement for innovation must increase.

Key lessons learned

A number of programmes and agencies are addressing the challenges mentioned above, reflected in an increasing number of procurement for innovations in Norway. However, a lesson learned so far is that it takes time to implement policies for procurement for innovation.

Measurement and impact assessment

There are several partial evaluation exercises, but no overall quantified targets nor a general evaluation encompassing all policy measures. The following partial assessments are relevant:

- Difi and the Supplier Development Programme have conducted two surveys on procurement for innovations. PWC conducts an annual sourcing survey with trends and findings from both private and public procurement. Innovation is one component in this study.
- The IRD PRD programme: In addition to annual customer surveys, the programme is being externally evaluated every fifth year. Latest evaluation in 2012.
- BI, Norwegian Business School, conducted in 2013 an in-depth study of the long term value creation based on the IRD/PRD projects.

There has no target been quantified yet for procurement for innovation; however, impact assessments, evaluation studies and studies of state of play are used.

Poland*

Strategic framework, action plan and scope for procurement for innovation policy

Poland does not have a stand-alone procurement for innovation action plan nor is the procurement for innovation action plan part of the country's general innovation or procurement strategy. The National Action Plan on Sustainable Public Procurement 2013-16 is a document in the process of implementation targeting strategic procurement, however its scope is limited to green and social procurements.

Procurement for innovation is regulated as part of the national procurement framework based on the Act of Public Procurement Law (PPL, 2004). The Public Procurement Law supports procurement of innovation based on the assumption that PPI requires some flexibility. Accordingly, the PPL includes provisions allowing the specification of functional and performance requirements in contract documents instead of recalling norms and standards; it provides possibility to conduct a technical dialog before launching the award procedure; for PPI projects it suggests to use PPI-friendly negotiated procedure or a procedure of competitive dialog; it provides for submitting variants or including innovative aspects in form of non-price award criteria.

Additionally, a 2014 amendment to the PPL provided for certain exclusions from the PPL scope to support research, experimentation, study or development which do not serve for mass production. Finally, the latest 2016 amendment implementing the new European directives introduced, among others, new procurement procedure of innovation partnership.

In addition to the PPL, the Strategy for innovation and efficiency of the economy "Dynamic Poland 2020" is relevant for procurement for innovation. The aim of the Dynamic Poland 2020 Strategy is to replace sectoral and administrative approaches to innovation policy with a horizontal and comprehensive perspective. Public procurement is one of the elements.

In line with the European legislation and with the aim of facilitation pre-commercial procurement is not covered by the PPL. The National Centre for Research and Development is involved in the organisational support and management of PCP projects. The Centre is an implementing agency performing tasks related to science, technology and innovation polices adopted by the Polish government. Its mission is to support the Polish research units and enterprises in developing their abilities to create and use solutions based on scientific research results in order to encourage economy development and to the benefit of society. The Centre manages the programmes and projects of scientific research and development in various sectors of the economy. In 2013 the Centre launched the project "Support for research and development through pre-commercial procurement". However, currently the project is suspended due to limited interest from the side of contracting authorities.

National references to PPI/PPC are based on the definitions provided for in the European legislation and guidelines.

* OECD Survey Part I submitted by Justyna Pożarowska, Counsellor General, Public Procurement Office.

Implementation

Poland undertakes the following specific actions to support procurement for innovation at the national level.

In 2008, the Council of Ministers adopted a document "The new approach to public procurement. Public Procurement and SMEs, Innovation and Sustainable Development". On the basis of this document the Public Procurement Office and the Polish Agency for Enterprise Development launched and implemented the "New Approach" project (2010-13). Activities performed within the framework of the project aimed at advancing the level of green, social, electronic and innovative procurement, as well as enhancing participation of SMEs as contractors in procurement procedures. The project component addressing PPI has been focused on trainings, consultancy and other knowledge dissemination activities. The outcome of the "New Approach" project included: 48 two-day trainings for local, regional and central level administrations awarding public contracts with 1 013 participants representing 646 institutions and entities; 3 conferences targeting all kinds of procurement market players; 3 conferences targeting auditors of Regional Chambers of Accounts as well as the Supreme Chamber of Control; elaboration of 28 publications and 19 journal articles.

Actions designed under the "New Approach" project are to be continued under the umbrella of current strategies.

Some further actions of relevance to the PPI/PCP are performed by the Polish Agency for Enterprise Development (PARP). PARP's general objective is to support sustainable economic growth in Poland. A more specific goal includes, among others, the creation of innovative eco-system across diverse innovation areas. The innovation ecosystem is supposed to involve companies, business support organisations, governmental unites as well as academia and other education entities. To this end, the PARP Agency is planning to develop the "inno_LAB" project with the aim to design and test new tools supporting innovation and leading to the increase in the capacity of national innovation system. Under the "inno_LAB" project some specific procurement projects are to be rendered with the use of PPI and PCP approach. For the time being, the project is in its initial stage of development.

Another project supporting innovative procurement from a more international angle is a joint initiative by PARP, the Ministry of Foreign Affairs and the Ministry of Development. In line with this project the Information Point for International Public Procurement has been launched to provide an information and access to global market of public procurement, including procurement systems of the European Union, the United Nations, the World Bank and other international institutions as well as of other countries. Except for information and advice, the Point is providing the space for networking (including partner searching exercises) potentially also in innovative procurement projects.

Furthermore, the Ministry of Development is governing the Smart Growth Operational Programme with some innovation-relevant targets in the component of "Intensification of co-operation within the framework of the national innovation system." Under this component, there is a space and funding to develop pilot projects building capacity of Polish enterprises in research, development and innovations. The PCP as well as PPI are indicated as forms of support expected to stimulate activity of national enterprises in R&D&I. Projects designed under this component are in their preparatory stages.

The PPI/PCP-related projects and programmes are generally financed from the National Strategies and Operational Programmes managing financial sources granted to Poland within the framework of the European New Financial Perspective. Specifically, the Operational Programme for Smart Growth supports PCP/PPI in public sector and Knowledge, Education and Development Operational Programme provides sources for social innovations as well as for trainings and workshops dedicated to awarding authorities/entities and audit institutions in respect of PPI.

Challenges and risks and solutions to overcome obstacles

In Poland, a more general problem for the national procurement market is a reluctance of contracting authorities to apply award criteria other than price. A challenge specific to innovative procurements is a higher level of risk attached to the innovation process.

To overcome these obstacles, the Public Procurement Office together with the Polish Agency for Enterprise Development are involved in the information/knowledge dissemination dedicated to the non-price criteria in public procurement. Additionally, the amendment of the Public Procurement Law radically limited the possibility of applying the price as the sole contract award criterion.

Key lessons learned

The adaptation of contracting entities to the above-mentioned change is ongoing.

Measurement, impact assessment

Poland has no system to measure the impact of actions related to procurement for innovation and no quantified targets have been established in respect of PPI. However, some assessments and evaluation studies regarding PPI have been elaborated (for more see www.paprp.gov.pl, information available mainly in Polish).

Since the procurement system in Poland is decentralised and individual awarding authorities independently apply the procurement law to award contracts responding to their needs, these authorities are the only holders of full and detailed information concerning awarded contracts (including PPI). While some data is collected and processed by the Public Procurement Office with the aim to provide general overview of national procurement market, specific data addressing PPI are not collected at the central level. Consequently, at the central level there is no system evaluating the impact of procurement for innovations. A basic statistic research based on samples of contract notices is conducted to assess the level of green and social procurement, however also without impact evaluation.

Portugal*

Strategic framework, action plan and scope for procurement for innovation policy

Portugal does not have a specific strategic framework for procurement for innovation or a stand-alone procurement for innovation action plan. Nevertheless, the general legal system in Portugal supports procurement for innovation. The most important regulation in Portugal's legal framework related to procurement is the Public Contracts Code (2008). The legal framework specifies the scope for procurement for innovation policy. This code makes e-procurement mandatory and is in this regard a motor of innovation. As a result of implementing e-procurement, SMEs (either alone or as part of an association) have better access to public markets because tender submission is easier. The code also sets the "most economically advantageous tender (MEAT)" criterion, which enables the contracting authority to consider criteria that reflect technical, innovative and sustainable aspects in addition to price. To facilitate access by SMEs, the code provides for measures like division into lots, adoption of regional criteria, and multi-access criteria.

Moreover, the Framework of ENEI (National Smart Specialization Strategy) gives particular relevance to public procurement as a demand-side instrument to drive and support innovation. Innovations related to environmental sustainability and the efficient use of resources are also targeted by the National Reform Programme (NRP) Green Tax Reform and the Green Growth Commitment (GGC), which was signed in 2015 by the Portuguese government and 82 institutions from the public sector, academic business and financial sector and from citizenship. The elements of the GGC are enabled by a set of catalysts, one being public procurement. The GGC mandates the establishment of a green public procurement programme; it also requires ensuring that sustainability criteria are included in all public procurement contracts.

Aside from general policy frameworks, innovation-related aspects also feature within the framework for public procurement. The Portuguese Public Procurement System (PPPS), managed by eSPap, IP, includes mandatory e-procurement for every procedure under eSPap's framework agreements (FA) (thus the dematerialisation of the tendering process), green criteria in the FA and, as applicable, the fostering of the participation of SME in the list of qualified suppliers to be part of each FA.

Portugal currently transposes the new European directives on public procurement into national law, which will result in significant changes to the Public Contracts Code, as well the GPPP - Green Public Procurement Policy.

In addition, Portugal developed the Roadmap for Eco-Innovation, which aims at fostering green growth through innovation; this roadmap is part of wider commitments and strategies related to green growth. One aim is to promote the competitiveness and internationalisation of Portugal's national economy.

* OECD Survey Part I submitted by Cristina Guedes, Head of division, General Directorate of Economic Activities.

Implementation

Portugal carried out the following specific actions and initiatives:

- dematerialisation of procurement procedures, from e-invitation to e-awarding - mandatory for every procedure under framework agreements regardless of value
- other PPPS-related e-tools (e-catalogue, e-reporting)
- green award criteria in most framework agreements (e.g. vehicles, paper, stationery, cleaning services, hardware, etc.)
- fostering the centralisation of procedures within PPPS.

The implementation of e-platforms, most notably the platform called "BASE", can be considered one of the most successful practices implemented to foster procurement for innovation in Portugal. The use of electronic procurement created an infrastructure that can be considered PPI friendly.

Portugal has transposed a number of EU policies to modernise public procurement.

On a project level, several European projects in the area of sustainable innovative procurement, were implemented. These projects focussed on achieving greater environmental efficiency and sustainability by using innovative procurement practices, and were sponsored by the European Union.

Portugal successfully implemented a number of innovative procurement processes. Goods and services procured range from the innovation of the electronic passport and automated passport screening, green innovations such as the reduction of paper records in hospitals or improved cleaning,

Most of Portugal's most successful practices are reflect in its procurement system (PPPS). The PPPS features the following core elements:

- e-procurement mandatory for every procedure
- green criteria
- fostering SME participation
- fostering aggregated procedures
- dematerialisation of procedures (namely via the related e-tools provided, such as e-catalogue, e-reporting, etc.).

Other good practices that have been demonstratively promoted procurement for innovation in Portugal, essentially correspond to general good procurement practices, namely to set a correct purchasing strategy, to develop the best specifications for the purpose, and to establish appropriate award criteria.

Challenges, risks and solutions to overcome obstacles

The new system caused disruption and required major changes in habits and attitudes, as well as the adjustment to the new legal framework, to e-Procurement and its set of tools. Significant challenges are:

- Costs of innovative solutions are high when compared to main stream solutions. This is relevant in moments of financial austerity.
- A zero risk culture that prevails with involved actors.
- Organisational barriers due to business-as-usual practices.
- Build the capacity of public procurement organisations to conduct specialised procurement types.
- Create dedicated units for innovation, environment, and financial questions in the public procurement organisations.
- Systematise evidences on the profit (life cycle costs) of innovative goods, services or works versus traditional ones.
- Develop financial mechanisms, tax breaks and other incentives to promote eco-procurement for innovation in strategic areas.
- Ensure that different electronic platforms are interoperable.

A thorough and ongoing communication plan helped overcome a number of challenges. The plan involved key stakeholders ranging from contracting authorities to Ministerial Purchasing Units to economic operators to other stakeholders. The following elements were implemented:

- a roadshow across Portugal, with over 1 200 participants
- an ongoing programme of training and capacity building sessions to key users
- annual conferences on public procurement
- presence of directors and officials in events, seminars and conferences procurement-related
- bilateral exchange of experiences with other European agencies, counterparties, Portuguese contracting authorities, Ministerial Purchasing Units
- themed working/technical groups.

Key lessons learned

A main lesson learned relates to communication, which is crucial to bring users on-board. The focus of the communication must be about the advantages of the system to the users as a way of better accepting the required changes. It is also very important to allow everyone to realise how they can contribute to improve the system, taking into account the legal and technological frameworks and their constraints and the major goals of the PPPS.

Measurement and impact assessment

A target for procurement for innovation has not yet been quantified. The impact of procurement for innovation is measured by impact assessments, evaluations studies and studies of state of play. A system to measure the impact of actions related to procurement for innovation was targeted through the National Green Public Procurement Action Plan, 2008-10, (ENCPE; note that ENCPE 2008-10 reached the end of its term; the follow-up plan was awaiting approval at the time of the OECD Survey) at the ENCPE monitoring report (May 2011)details results of the results of ENCPE. This report found that "in 2010, over 56% of the procedures and over 60% of the total value of acquisitions of goods and services classified in priority categories [...] incorporated environmental criteria or requirements." A National Survey on e-Procurement was conducted to assess the impact of electronic public procurement after one year of mandatory use in Portugal. eSPap calculates the percentage of SMEs based on data gathered in the e-portal BASE. Every public contracting procedure must be published in this database.

Russian Federation*

Strategic framework, action plan and scope for procurement for innovation policy

The Russian Federation specifies requirements related to procurement for innovation in the law, including obligations (as percentage shares) for innovation products to be procured. State-owned enterprises (SOEs) are obliged to purchase innovations and to publish their plans for procurement for innovation. Initially, the target is set at 2.5%; this target will be increased to 5%. The government will issue regulations on how to calculate the target. Than the plans for procurement for innovation should be published by the middle of the next year. An official must be named in every SOE who is in charge of innovation programmes and procurement. The value for innovation might be 20-25% (under discussion).

In addition, the Russian Federation has a stand-alone procurement for innovation action plan, but experience in working with innovative contracts is lacking. The scope is not based on existing definitions; in the Russian Federation, procurement for innovation comprises procurement of innovative goods, services, technology.

Implementation

Among the specific actions implemented in the Russian Federation to support procurement for innovation are legal obligations control and one-off cases.

Challenges, risks and solutions to overcome obstacles

The main challenges in the Russian Federation that hinder procurement for innovation are related to the formal approach often taken by administrations. There is also a lack of interest, administrative barriers, inadequate specifications related to public funds, and restrictions in the legislation. The Russian Federation has yet to overcome these obstacles, with the exception of public finance restrictions.

Key lessons learned

The Russian Federation's key lesson learned is that innovative procurement can be implemented in innovation-friendly environment.

Measurement and impact assessment

The Russian Federation does not yet have a system in place to measure the impact of actions related to procurement for innovation. As mentioned above, the Russian Federation has quantified targets for procurement for innovation and follows up on activities to reach of this target. The targets have not yet been achieved. Because of the formal approach taken in the administration, there are no impact assessments, evaluation studies and/or studies of state of play regarding any type of procurement for innovation.

* OECD Survey Part I submitted by Olga Anchishkina, head of the board, NAPE.

Serbia[*]

Strategic framework, action plan and scope for procurement for innovation policy

Considering that there is no strategic framework on procurement for innovation in Serbia, the country follows on World Bank procedures, slightly modified and simplified to be more tailored to the best practice and well-established private sector procurement methods. These amended procedures (i.e. the commercial practice) are described in a manual.

The scope of procurement for innovation policy is in line with the World Bank procurement guidelines. The Project Implementation Document (or Manual) describes the basic guiding principles and acceptable procedures applicable to the loan. For example, there are mandatory provisions that beneficiaries of the loan shall not award contracts to their parent or affiliate companies unless there is an established arms-length arrangement. Also, the following needs are to be determined:

- assessing the capacity of the beneficiaries to carry out procurement efficiently
- approving acceptable plans for the procurement of goods, works, and non-consulting services, and the selection of consultants
- agreeing to supervision and oversight arrangements for the procurement to be carried out by the beneficiaries so as to ensure compliance with the agreed private sector methods and commercial practices
- maintaining all relevant records for the World Bank's post review and audits when requested.

Implementation

There are no specific actions to support procurement for innovation at the national/central/federal level in Serbia. There have been initiatives which were implemented and supervised by the Innovation Fund (IF). However, these initiatives are related to the IF activities only.

Challenges, risks and solutions to overcome obstacles

A major challenge in Serbia was contract management. This challenge has been overcome to some extent.

Key lessons learned

A key lesson learned in Serbia relates to the importance of good timing and reviewing of deliverables, as well as the inspection of goods delivered.

Measurement and impact assessment

A system is not in place to measure the impact of actions related to procurement for innovation.

[*] OECD Survey Part I submitted by Ljiljana Krejovic, Procurement Specialist, Inovation Fund.

Slovak Republic*

Strategic framework, action plan and scope for procurement for innovation policy

The Slovak Republic does not have a strategic framework yet, and there is no stand-alone action plan. However, there are initiatives in support of procurement for innovation planned.

As a central state body responsible for supporting innovations, the Ministry of Economy of the Slovak Republic is currently preparing a national project with the aim to motivate public sector entities to purchase innovative services, products and construction works. Together with other relevant central state bodies (Ministry of Interior, Ministry of Finance, Office for Public Procurement, etc.) the Ministry of the Economy intends to define a public procurement methodology aimed at supporting the public sector in choosing the best contractor taking into account the innovative aspect and added value of the proposed solution (by introducing other criteria for the evaluation of public procurement applications such as life-cycle costs, overall performance, environmental impact and impact on society, etc.), identify potential innovations for the public authorities' purchases or for offering incentives by public authorities and spread good practice cases. These measures will establish a Slovak procurement for innovation strategy. In doing so, the Slovak Republic will use the definitions of PCP and PPI from the European Commission as defined in the rules for participation of the European Framework Programme definitions to frame the scope for procurement for innovation policy.

* OECD Survey Part I submitted by Lucia Prozbikova, Senior state advisor, Ministry of Economy.

Spain*

Strategic framework, action plan and scope for procurement for innovation policy

Spain does not have a stand-alone procurement for innovation action plan, the procurement for innovation action plan is both part of the country's general innovation strategy and part of the procurement strategy:

The innovation strategy was adopted in 2010 and includes specific support to procurement for innovation in particular lead markets (i.e. Health, Energy, e-Administration, Defence, etc.). A regulation on procurement for innovation complements the strategy; this regulation was passed in 2011. This regulation obliges all ministries to specify the amounts allocated to procurement for innovation in their budgets and in different multiannual action programmes. A target of 3% in new investment for procurement for innovation was pursued for the year 2013.

In addition, procurement for innovation is supported by the Spanish Strategy for Science, Technology and Innovation 2013-20 and the more specific and operative Plan for Scientific, Technical and Innovation Research 2013-16. The Royal Decree 345/2012 tasked the Directorate General of Innovation and Competitiveness (DGIC) with the promotion and dissemination on the use of procurement for innovation.

Funding for procurement for innovation comes from structural funds, as well as specific allocations in budgets. The current Programme 2014-20 allots EUR 300 million within the pluri-regional programme to procurement for innovation. To reach the target of 3%, the Ministry of Economy and Competitiveness (MINECO) and the Centre for the Development of Industrial Technology (CDTI) developed a set of elements to promote procurement for innovation and support the implementation of the legal framework. A Guide on Innovative Public Procurement (first version 2011, update December 2015) describes administrative actions to promote innovation by focussing on the demand-side and the creation of markets. The guide foresees two complementary mechanisms:

1. Pre-commercial procurement: "Compra Pública Pre commercial (CPP)", which purchases new solutions and includes risk-benefit sharing between supplier and procurer.
2. Innovative technology public procurement: "Compra Pública de Tecnología Innovadora (CPTI)", i.e., the procurement of commercial – not yet existing -- mature solutions that could be developed in a reasonable time. In addition, the National Institute for Public Administrations (INAP) published a book on the foundations and instrumentation of IP, aimed at wide spreading the use of this tool, in 2013.

An example for specific actions to support procurement for innovation at the national level is the INNODEMANDA programme. This instrument provides financing and is managed by the Centre for the Development of Industrial Technology. Funds will be given to companies to cover innovation costs so that the public body pays the same

* OECD Survey Part I submitted by Miguel Ortiz, Innovation Procurement Coordinator, CDTI.

amount as if it bought the already developed technology, in this way, affording greater exposure of these companies' products and services in the administration.

Another example is the INNOCOMPRA programme, implemented through FID (Fostering Innovation through Demand) Agreements. This programme is managed by the MINECO and uses EU Structural Funds, ERDF, to co-finance procurement for innovations at regional level. Until July 2014, 21 operations had been covered by this instrument, mobilising EUR 230 million. At the end of 2015, those 21 operations were thoroughly monitored and audited, finding a high degree of technical fulfillment with something less than 90% of technical contains been completed.

These two examples (INNODEMANDA and INNOCOMPRA) encouraged more than 25 procurement for innovations within Spain, either CPP or CPTI.

Implementation

Since 2011, Spain has developed a programme for co-financing actions on PPI with support of the EU - FEDER Technological Fund. During the last budget period 2007-13, 21 actions have been developed for an amount of EUR 230 million.

Since 2014, Spain is preparing a new set of PPI proposals for a global amount of EUR 300 million to be co-financed with this FEDER Technological Fund through the new Spanish Programme on PPI (FID) for 2014-20.

It is highlighted the Programme FID SALUD that aims at systematically improving public health services portfolio through yearly calls of PPI. Currently, on its second year of existence, the programme involved every regional health service (18, including Ceuta and Melilla) and is technically co-ordinated by the Health Ministry in order to prevent duplication and to foster synergies. Structural financing and oversight is provided by the Ministry of Economy and Competitiveness. To date more than 40 proposals have been independently assessed by ISCIII (Health Institute Carlos III) and 15 have been approved mobilising some EUR 62 million just for the 2015 call.

Spain considered the following among its most successful practices:

- Projects Hospital 2050 and INNOVASAUDE (www.sergas.es/Hospital-2050--Innova-Saude?idioma=es) for the period 2007-13 and several proposals for the new period included in FID SALUD. Hospital 2050 is addressed to health innovation for smart general management and patients management systems, traceability, robotisation, hospital buildings sustainability, clinical information security and so on.
- Initiatives on aeronautical technologies come from the 2007-13 period and continue to be relevant during the current period with the Civil UAVs (Unmanned Aerial Vehicles) Initiative (http://gain.xunta.es/artigos/466/civil+uavs+initiative). Targets of this last one consists in developing a new offer of public services for forests protection, land management, civil protection against catastrophes , coasts surveillance and so on. Also, it creates a technological and industrial park for UAVs.
- R&D and Innovation for the rural environment. RECUPERA 2020 (www.recupera2020.csic.es/en) is a collaborative project between the Spanish National Scientific Research Council, other public research centres and companies.

- Several projects during 2007-13 on implementing innovations for smart cities (SMART Coruña www.coruna.gal/servlet/Satellite?c=Page&d=&cid=1377134531332&pagename=Smart/Page/Generico-Page-Generica), entrepreneurship centres (ALCALA+i www.laprocesadora.com), etc.
- New initiatives for 2014-20 of other ministries for motorcycles and cars safety in roads and for improvement of meteorological fog conditions in motorways.

Challenges, risks and solutions to overcome the obstacles

Spain experienced the following challenges:

- Spain implemented intensive promotion campaigns within the national administration and in most of the regions in Spain. However, there is still significant room for increasing awareness within Spanish procurers regarding the benefits and possibilities of IP, the new legal framework and financial instrument that can support it.
- Administrative ranks are generally risk averse when it comes to implementing new procurement procedures. This also applies to auditors, sometimes due to new legal framework uncertainties.
- Public procurers conduct too few analyses of requirements and open market consultations. There is a need for more promotion and acknowledgement to incentivise continued education.
- Lack of money is an obstacle to assume higher risk in procurements: public procurers can afford to address urgent needs only.
- Procurers are new stakeholders in Horizon 2020, EU Research and Innovation Programme, and funding is not so attractive (management costs vs. procurement costs).

Many of the mentioned obstacles are not in the hand of government an administration to be overcome.

Key lessons learned

Spain reported the following observations:

- Co-ordination between national and regional administrations is essential, either for promotion of procurement for innovation or for complementing financial support mechanisms (EU Structural Funds, ERDF).
- Mappings of early demand, published by public procurers and targeted at potential suppliers would help to boost innovation by highlighting needs. Several of these demand mappings for different markets were already published in Spain; however, it is difficult to retrieve the mappings from the public procurers that develop the mapping.
- Easing the scope of structural funds set aside for R&D&I, in the sense of permitting the entry into service of procured innovation on a pilot scale beyond the strict definition of innovation, will certainly ease the general adoption of procurement for innovation tools by the community of public procurers.

- The European Commission should facilitate the syndication of different funding sources, i.e funding sources should be co-ordinated and complementary. This is important to fully cover the process from the research and development phase to entry into service (EIS) at a wider scale. By doing so, support to procurement for innovation would fully tap into potential anticipated by econometric models modelling the demand for procurement for innovation.
- Innovation fields covered reach mainly to health, ICTs and aeronautics. New actions should be implemented to extend broadly those three fields and to reach new ones yet not well covered, such as energy, environmental, surveillance and security technologies and innovations.
- Increase and improve participation of innovation procurers coming from ministries and from regional and local authorities. Also coming from universities.
- Improve training of public procurers, making them aware to practices and advantages of procurement for innovation.

Measurement and impact assessment

Spain tracks its quantitative target for procurement for innovation through a marker in the State Procurement Platform (PACE). This marker indicates if a project corresponds to IP. There is currently little follow up if the target is fulfilled, apart from the evaluation as part of the INNODEMANDA/INNOCOMPRA programmes, as well as otherwise un-evaluated procurement for innovations entered in PACE. More resources could be focused on this particular matter.

One of the projects tracked by the procurement for innovation marker in PACE that is not part of the INNODEMANDA / INNOCOMPRA programme is a new investment project called "Type K" which is in use since 2013.

Sweden*

Strategic framework, action plan and scope for procurement for innovation policy

Sweden does not have a specific procurement for innovation action plan. Instead, procurement for innovation is embedded into the National Public Procurement Strategy (2016). The public procurement strategy is mostly directed to governmental agencies since regional and local levels of government are independent in Sweden. The strategy does however give guidance on a wide variety of procurement for innovation aspects to national, regional and local levels of government as well as suppliers. It was also embedded into the Swedish innovation strategy (2012). Some regions, county councils and larger municipalities also include procurement for innovation in their strategies on development, innovation and procurement. Other elements supporting procurement for innovation have been included in other policy documents, such as *smart industry* the strategy for new industrialisation, where promotion of procurement for innovation is included in the action plan. Since regional and local levels of government have a high level of independency in Sweden, many of the strategic decisions to pursue procurement for innovation are taken at the regional or local level. Some regions, county councils and larger municipalities include procurement for innovation in their strategies for innovation and public procurement. The National Agency for Public Procurement, together with Sweden's innovation Agency VINNOVA and other agencies, defined the scope for procurement for innovation, including new definitions corresponding to the definitions used by the European Commission and in line with ERAC definitions. That means that Sweden includes procurements that open up the possibility for suppliers to submit tenders with innovative solutions.

Implementation

The following institutions are the main agencies supporting Sweden's procurement for innovation efforts:

- The National Agency for Public Procurement took over all responsibility for providing support regardingpublic procurement from the Swedish Competition Authority in September 2015. The agency provides support, guidance and information on all aspects of procurement for innovation under its general procurement support activities. The agency provides guidelines, collects and disseminates good examples, and provides methodological support to specific procurement for innovation projects. The agency also reaches out to other authorities/organisations in Sweden as well as internationally.
- VINNOVA - Sweden's Innovation Agency offers financing possibilities to contracting authorities that wish to procure innovation. Target groups are municipalities, county councils, regions, central government authorities and other contracting authorities. Lessons learned from the supported projects are gathered jointly by VINNOVA and the National Agency for Public Procurement. The two

* OECD Survey Part I submitted by Lena Svendsen, Head of Section , Ministry of Enterprise and Innovation.

agencies have also signed a mutual agreement to co-operate and find synergies in the area of procurement for innovation. Financing for procurement for innovation was first made available by VINNOVA in 2011.

- The Swedish Energy Agency has initiated, co-founded and participated in nearly 60 different technology procurements (catalytic procurement), with the aim of accelerating the development of energy-efficient products in the early 1990s. These activities are on-going and expanding. The energy agency has co-operated with VINNOVA on public technology and procurement for innovation related to environment technology since 2012.
- The Swedish Transport Administration analyses where in its operations it can use procurement for innovation to support development of effective processes and technologies since 2012. The analysis is based on previous procurements, and will result in a plan for increased procurement for innovation within the Transport Administration. The long-term aim is to increase efficiency and usefulness within the Administration's operations and thereby contribute to an economically efficient and long-term sustainable transport system.
- The Swedish Association of Local Authorities and Regions (SALAR) is both an employers' organisation and an organisation that represents and advocates for local government in Sweden. All of Sweden's municipalities, county councils and regions are members of SALAR. SALAR performs activities to encourage procurement for innovation among its members, such as publishing inspirational material and leading knowledge development. Currently, SALAR is researching areas for future action involving buyer's groups.

The following activities are considered particularly useful in Sweden.

Policy

In 2014, several central government authorities were tasked (via appropriation directions) to analyse their need for development and innovation within their areas of responsibility. The authorities were also asked to identify possible ways to solve these needs, including procurement for innovation. Results were reported in the beginning of 2015. The measure has led to authorities gaining a deeper understanding of the possibilities of procurement for innovation.

Programmes/financial instruments

1. VINNOVA launched a programme to finance procurements of innovation in 2011. The aim was to encourage Swedish contracting authorities to perform procurements of innovation and gain experience. Up until now, some 35 projects have been financed, mostly pre-studies and pre-commercial procurements, and knowledge about procurement of innovation in Sweden has increased. The most important element is a specification which requires the contracting authorities to perform the procurements themselves (VINNOVA does not do it on their behalf). This encourages learning within the organisations. In addition, authorities are required to contribute at least half of the project costs themselves. This means that the projects are needs-driven. The managers of all projects have an opportunity to meet and learn from each other, which ensures mutual learning and dissemination. This project highlighted how external financing consultants can accelerate the

process. While such financing consultants are valuable as project managers or experts, it is vital that the project is grounded within the contracting agency and that the agency's employees (including procurers and development personnel) contribute substantial efforts.

2. The catalytic procurements of the Swedish Energy Agency have a history back to early 1990. The Energy Agency does not perform procurements itself, but finances and facilitates buyers groups with common needs within specific areas (for example owners of housing or office buildings). Buyers groups can be comprised of both contracting authorities and private companies. There is evidence to suggest that the initiative constitutes best practice, as good effects have been demonstrated on many of the more than 60 procurements of technology that has been performed since early 1990. Buyers groups may be of particular importance in counties such as Sweden, where there are many small, but independent, contracting authorities.
3. The National Agency for Public Procurement launched in the beginning of 2016 a programme with the aim to develop public procurement with regards to innovation and dialogue with the market. The programme will run until 2019 and focus mostly on support for the phase before procurement for innovation: identifying and analysing needs as well as early dialogue with the market. The Programme have three different focus areas: a) methodological support to specific procurement for innovation projects; b) Collect and disseminate knowledge in form of guidelines and good examples; c) facilitate networking and experience sharing among contracting authorities.

Best practice stand-alone cases

Experiences and lessons learned from cases have been identified as a good way to both motivate and support contracting authorities. This aims at both motivating authorities to learn about procurement for innovation, as well as implement it. The National Agency for Public Procurement has identified examples and lessons learned from procurement for innovation projects, as well as identifying what kind of support is required in order to assist the contracting authorities to apply procurement for innovation in future.

Challenges, risks and solutions to overcome obstacles

Procurement for innovation policy and initiatives in Sweden must take into account the fragmented demand of Sweden's independent regional and local contracting authorities, which represent the bulk of the public demand in Sweden. This gives rise to two challenges:

- Swedish regional and local public organisations cannot be ordered to use procurement for innovation. Policy measures must focus on encouraging contracting authorities to consider procurement for innovation as a means to improving their performance of public services (both short and long term).
- Some contracting authorities are too small to be able to undertake procurement for innovation. Therefore, policies need to provide solutions for co-ordinating similar demand in separate organisations (e.g. through buyers groups, as does the Swedish Energy Agency).

Another challenge is encouraging contracting authorities to define their challenges and procurement needs in a strategic context.

Even with needs and challenges identified, verified, prioritised and co-ordinated, obstacles remain for widespread use of procurement for innovation. These obstacles include perceived legal obstacles and risks (real and imagined), lack of governance and management, myopic focus and lack of time, attitudes and habits within the public administration and inflexible funding schemes. Sweden has overcome these obstacles partly, but not completely. A lot still remains to be done.

Key lessons learned

One lesson learned is that the implementation of procurement for innovation is a slow process that can only be implemented gradually. Procurement for innovation must be based on providing real benefits for contracting authorities. Policy makers can take this into account by setting up initiatives to increase knowledge on procurement for innovation practice, provide guidance and offer financing opportunities.

Measurement and impact assessment

Sweden does not yet have a system to measure the collective impact of procurement for innovation-related actions, but the country does use evaluation tools to measure results, outcomes and impacts of individual procurement for innovation initiatives. These are followed up by the authorities/organisations in charge of the respective initiative, sometimes in co-operation with other bodies such as the Swedish Agency for Growth Policy Analysis. Since most of the initiatives are fairly recent (with the exception of the Swedish Energy Agency), these activities mainly concern specific cases.

There is no quantified target for procurement for innovation on a national level, due to statistical difficulties. It can be questioned whether a quantified national target is useful in the Swedish case, given that the vast majority of public procurements are carried out by independent regional and local bodies, which would not be bound by a national target. Sub-dividing an overall target would be complicated. Aside from quantitative targets, introducing qualitative targets may currently be equally or more important than quantitative. "Soft" measures to increase knowledge and co-ordinating demand for procurement for innovation are deemed more effective at this point. However, this may change in the future.

Switzerland*

Strategic framework, action plan and scope for procurement for innovation policy

Switzerland does not have a specific strategic framework for procurement for innovation nor a stand-alone procurement for innovation action plan.

Regarding research and innovation in Switzerland, the bottom-up approach dominates. Moreover, the Federal Council applies an integrated approach with regard to innovation, combing the promotion of education, research and innovation in one federal act with new legislation. The new State Secretariat for Education, Research and Innovation (SERI) defines the overall (legal) framework for innovation support, whereas the Commission for Technology and Innovation (CTI) is the Swiss innovation promotion agency that encourages scientific innovation in Switzerland by providing funding, advisory services and networks to improve the Swiss economy. Support is based on the principle of subsidiarity, i.e. only stepping in when innovation and market opportunities would remain untapped without the CTI's support. While Swiss businesses undertake and finance a large part of their research and innovation (R&I) activities themselves, the total amount of public funding for innovation activities is much lower than for basic research and is only available for publicly-oriented R&I partners, such as public universities or research and technology organisations carrying out applied research with or for Swiss enterprises.

With this policy, Switzerland views itself as an exception among the member countries of the OECD. Switzerland's innovation policy assumes that the generation of innovation is a core task of industry and SMEs and that the private sector has to contribute own effort and money to a large extent. Consequently, the Swiss government only rarely intervenes directly in the innovation process. Instead, it focuses on setting the legal framework and providing the necessary infrastructure (i.e. the adequate regulatory framework for the actors involved in innovation, mainly from industry).

Implementation

With regard to specific actions supporting procurement for innovation in Switzerland, there is - apart from funding R&I projects - a strong emphasis on mentoring, coaching and teaching business knowledge and entrepreneurship skills as well as providing small companies and start-ups with a network of partners in the R&I sector as well as along their value chain. Support is also available for export and internationalisation.

As pointed out, Switzerland is not inclined towards a strategic framework for procurement for innovation. However, for the time being, the following measures can be considered as aiming at promoting innovation in general (besides the above mentioned measures by CTI):

- The SwissEnergy programme encompasses the promotion of progressive projects relating to energy efficiency and the use of renewable energy. In 2011, the

* OECD Survey Part I submitted by Myriam Cevallos, Scientific advisor, State Secretariat for Education, Research and Innovation.

programme entered its third decade of activity. The decision by the Federal Council to withdraw from the use of nuclear energy on a step-by-step basis means that the objectives of SwissEnergy will grow in importance, and the programme will therefore play a significant role in the restructuring of Switzerland's energy supply in the coming decades. The Swiss Federal Office of Energy (SFOE) is responsible for the operational management of the programme.

- The National Research Programmes (NRP) was established in in 1975. Since, the NRP's mission has been to generate scientific knowledge aimed at solving Switzerland's most pressing problems. The Federal Council specifies the research topics of the individual programmes. NRPs are contributing scientifically to the solution of these problems, for example by developing action plans, providing political advice and creating special research infrastructures. NRP 62 "Smart Materials' initiative is a good example of the routes that research can take, from the initial idea through to product development. The Federal Council usually commissions two to four NRPs at a time with a budget of 10 to 15 million Swiss francs per project. Proposals are evaluated by the Swiss National Science Foundation (SNSF).
- The Swiss Competence Centres for Energy Research (SCCER) are contributing to innovation in the area of renewable energy. In order to implement the decision by the Federal Council and parliament to phase out nuclear power by 2035, the efficiency of renewable energies must be increased and new opportunities in this area must be exploited. The Federal Council intends to boost research into renewable energy under the Coordinated Energy Research in Switzerland Action Plan. The key element in this plan involves establishing inter-university research networks, or Swiss Competence Centres for Energy Research (SCCER). This task was given to the CTI, with the support of the Swiss National Science Foundation.

Challenges, risks and solutions to overcome obstacles

A recent report on the Swiss innovation policy dating from 2013 (*Franz Barjak, Wirkungen innovationspolitischer Foerdermassnahmen in der Schweiz, Studie im Auftrag des Staatssekretariats fuer Bildung, Forschung und Innovation*, 2013) highlights the lack of co-ordination and co-operation between the various stakeholders as a shortcoming of Swiss innovation policy. Moreover, the measures mentioned above were launched as singular measures without integrating them in an overall innovation strategy.

Measurement and impact assessment

As procurement for innovation is not at the centre of Swiss innovation policy, there is no measurement of the impacts of actions related to procurement for innovation nor does the government or other institutions define targets with regard to procurement for innovation. However, the above mentioned recent report on Swiss innovation policy dating from 2013 comes to the conclusion that no fundamental changes to the support measures are required as the Swiss system proves to be generally adequate and consistent, efficiently implemented, effective and to have a positive impact on technical progress. Yet, the report also highlights some shortcomings, especially with regard to the "culture of support" that constitutes a mental barrier to new approaches among companies and academic institutions.

Turkey*

Strategic framework, action plan and scope for procurement for innovation policy

Tukey's "Programme for Technology Development and Domestic Production through Public Procurement" is one of the 25 primary transformation programmes within the frame of 10th National Development Plan (2014-18). The Grand National Assembly of Turkey approved the programme in 2013. The aim of the programme is to use public procurement to promote innovation, domestic production, technology transfer and innovative entrepreneurship. The programme's scope comprises several aspects: considering a domestic R&D and innovation contribution requirement in public procurement and establish a right to use allocations; promoting innovation, domestic industry and technology transfer; and increasing foreign direct investment (FDI) by policies implemented in public procurement. The programme is co-ordinated by the Ministry of Science, Industry and Technology.

After the adoption of the Tenth Development Plan and Priority Transformation Programs by the Grand National Assembly of Turkey, a (stand-alone) action plan for "Program for Technology Development through Public Procurement" has been prepared under the co-ordination of Ministry of Science, Industry and Technology. The action plan consists of 17 policy issues under 5 main components and 17 actions assigned to a consortium of ministries, public bodies, NGOs and chambers of industry.

This procurement for innovation action plan is not part of the country's general innovation or procurement strategy. The "Program for Technology Development through Public Procurement" is part of the 10th Development Plan, but not part of the National STI Strategy 2011-16 or part of the strategy on public procurement.

The "Program for Technology Development and Domestic Production through Public Procurement" within the frame of 10th National Development Plan (2014-18) has the following concrete targets:

- increasing the share of domestic firms in medium-high and high technology sectors in public procurement
- supporting international branding in high-technology sectors and increasing the number of branded products
- increasing R&D expenditure by means of the public procurement system
- increasing FDI by policies that will be implemented in public procurement.

The performance indicators to monitor these targets include the following:

- share of domestic production in public procurement
- share of SMEs in public procurement
- share of domestic production based on public purchase guarantee in total public procurement

* OECD Survey Part I submitted by Yasemin Aslan, Director, Department of STI Policy, TUBITAK.

- in procurements from abroad: share of domestic firms established with offset practices
- number and amount of offset agreements.

The programme contains the following components:

- reorganising the public procurement system in a way that promotes R&D and innovation
- developing a financing and organisation model
- developing a legislative infrastructure
- increasing entrepreneurship and innovation capacity of the private sector by using public procurement
- developing institutional capacity.

Implementation

Specific actions: Aside from the "Program for Technology Development and Domestic Production Through Public Procurement" within the frame of 10th National Development Plan (2014-18), there have recently been further efforts to enhance innovative capacity by procurement.

The Supreme Council for Science and Technology (SCST), which is the highest policy making body for STI issues, since 2011 has issued decrees calling for "'improved public procurement mechanisms', 'licensing to foster innovation, localisation and technology transfer', 'more digital content in education' and 'greater use of technology in government service delivery'."

To foster R&D-based procurement methods, Turkey launched the "Public Institutions Research and Development Projects Support Program" (TUBITAK-1007) in 2005. The programme has been designed to fulfil the R&D needs of public institutions via dedicated calls for R&D projects by universities, industry and public research institutes. R&D-based requirements of the public sector are met by means of results-oriented R&D projects with no budgetary limitations.

Best practices have yet to materialise, given that the programme to support procurement for innovation and its action plan have only recently been launched.

Challenges, risks and solutions to overcome obstacles

The co-ordination between stakeholders, public-public and public-private sector interaction; as well as the need for improvement in the legislative framework regarding the public procurement and right of use allocations in a way that will promote innovation and technology transfer was few of the major challenges.

The co-ordination challenge was overcome by inclusive workshops with the stakeholders and strong dialogue between high level officials.

Key lessons learned

Turkey considered a key lesson learned that inclusive policy-making process, engaging all related parties, leads to a more accepted strategy are necessary. Support from high policy levels enables the implementation of the strategy to be more thorough.

Measurement and impact assessment

The Tenth Development Plan, with a strategic perspective at its core, covers economic, social, sectoral and regional areas, as well as setting forth the critical priority areas of intervention through its priority transformation programmes. Programmes are composed of public policies for priority areas that cover more than one sector and facilitate monitoring and implementation of plans. Programme details, sub-components, implementation activities and projects, budget requirements and legislative infrastructure have been transformed into action plans with joint participation and contribution of co-ordinator and responsible institutions for the components. The Ministry of Development monitors the progress through "Programme Monitoring Reports" which are due every six months. The High Planning Council is the authority with the right of revision of the programmes if needed, considering the implementation results.

The impact of the "Programme for Technology Development and Domestic Production through Public Procurement" is monitored by performance indicators, which are defined during the development stage (see above "Strategic framework".) The targets have not yet been achieved.

There are no impact assessments, evaluation studies and/or studies of state of play regarding any type of procurement for innovation.

United Kingdom*

Strategic framework, action plan and scope for procurement for innovation policy

The United Kingdom's main vehicle for taking forward procurement for innovation is the Small Business Research Initiative (SBRI).

SBRI is a competition-based innovation programme managed by Innovate UK, the United Kingdom's innovation agency that provides opportunities for companies to engage with the public sector to develop and provide new products and services for policy and operational challenges. SBRI provides 100% R&D funding to support companies to develop solutions. The intellectual property rights remain with the company, which is then able to market the product commercially more widely.

The United Kingdom's SBRI was established in its current structure in 2009. It is closely modelled on the US Small Business Innovation Research programme, which was introduced in 1982, and is generally considered to be the leading model for public procurement for innovation from SMEs.

The procurement for innovation action plan is part of the country's general innovation or procurement strategy. Procurement for innovation and the SBRI programme was included in the United Kingdom's Science and Innovation Strategy of the previous 2010 to 2015 Conservative and Liberal Democrat coalition government in December. There is no current published innovation strategy from the new government administration (from May 2015).

SBRI is run under EU rules for Pre-Commercial Procurement. These were codified in 2006 in order to enable member states to adopt programmes like the US SBIR.

The SBRI programme is considered to be a policy mechanism that supports technological development along a sequence of steps which ends with the commercial (i.e. on the market) availability of a product or service. The SBRI programme can be mapped to part of a Procurement of Innovation Framework, used extensively by the European Union. Organisations of all sizes and from anywhere in the European Union are eligible to apply for UK SBRI contracts.

Implementation

The SBRI programme is supported by Innovate UK, the United Kingdom's innovation agency, and can be used by government departments and public bodies.

Challenges, risks and solutions to overcome obstacles

For implementing SBRI, the most significant challenges have been:

- Ensuring awareness about the programme and its potential benefits across departments and public bodies.

* OECD Survey Part I submitted by Stuart Barthropp, Assistant Director - Innovation Procurement, Department for Business Innovation and Skills.

- Perceptions of SBRI as complex or risky compared to traditional procurement methods.
- Departments and public bodies needing to find funding for SBRI competitions. There is no central funding, and departments need to fund their own SBRI competitions.

The United Kingdom is still working to tackle these challenges. For example, Innovate UK has established an SBRI Practitioners Community of Practice which provides a forum to share best practice across government departments. SBRI has been growing steadily since it was relaunched in its current format in April 2009, with the value of contracts awarded through the programme rising from GBP 13 million in 2010/11 to GBP 83 million in 2014/15. Overall, SBRI has provided businesses with over GBP 270 million of contracts since 2009. There are now over 70 departments and agencies that have used the programme.

Key lessons learned

The United Kingdom highlighted the following most important lesson learned:

- Senior level leadership is needed in departments to drive the use of SBRI.
- Funding for SBRI in departments can be difficult, for example with those wishing to use SBRI having no access to relevant budgets.
- There need to be a constant flow of case studies and evidence to show the benefits of SBRI in order to persuade more departments to adopt it.

Measurement and impact assessment

Innovate UK have commissioned an independent evaluation of the SBRI programme. The terms of reference for the evaluation are to:

- draw up a baseline for SBRI and to understand the effect of the new targets
- review the SBRI process
- review the impact of the programme.

There is no quantified target for procurement for innovation, but the United Kingdom attempts to measure and follow up activities to reach (qualitative) targets.

Innovate UK's commissioned evaluation report for SBRI will be published shortly.

United States*

Strategic framework, action plan and scope for procurement for innovation policy

The United States has a stand-alone action plan on procurement for innovation, issued by the Office of Management and Budget in 2010, and titled "25 Point Implementation Plan to Reform Federal Information Technology Management". The plan contains a number of quantified targets, such as: terminating or turning around at least one-third of underperforming projects in the IT portfolio; increasing cloud usage; reducing the number of federal data centres by at least 800 by 2015. In addition, the plan envisions to solidify and consolidate funding, introduce flexible budget models, increasing the professional capacity, and launching an interactive platform for agency/industry collaboration.

One of the US procurement initiatives focuses on increasing innovation in technology and streamlining procurement. It includes the following elements:

- Smarter IT delivery by buying strategically and by buying in phases (modular contracting, agile purchases).
- TechFAR: guidance for using innovative practices to buy IT.
- Open dialogue summary.
- Issuing Streamlining Procurement Management with the initiative: Transforming the Marketplace: Simplifying Federal Procurement to Improve Performance, Drive Innovation, and Increase Savings (2014).
- Procurement for innovation items are also included in the budget.

Implementation

The United States has a number of initiatives underway to support the use of innovative procurement practices. Initiatives include new tools that help agencies adopt industry best practices and new talent development models that give contracting officials hands-on experience applying these tools with the help of experts. They also include initiatives to simplify procedures, include feedback from external and internal stakeholders, and dispel "myths" that may discourage the workforce from engaging with vendors as they plan and undertake acquisitions.

The "TechFAR" handbook is considered one of the best practices in the United States. The TechFAR provides agency personnel involved in the procurement process with practical tools for applying industry best practices to digital services acquisitions. Specifically, the TechFAR discusses relevant FAR authorities and includes practice tips and sample language. It is a living document. All federal agency stakeholders, including representatives from contracting, the programme office, Chief Information Officers, and Office of General Counsel, are encouraged to use this guidance. Readers are urged to provide feedback, share experiences, and offer additional strategies or practice tips that might be used to assure that IT acquisitions achieve their desired results. This current edition of TechFAR focuses on how to use contractors in iterative development

* OECD Survey Part I submitted by Julia Wise, Procurement Policy Analyst, OMB.

processes. Specifically, TechFAR includes provisions that are relevant to a commercial methodology named "Agile Development". This methodology is characterised by incremental and iterative processes where products are developed in close collaboration with customers. Agile development is geared towards projects where significant design and development are needed, such as citizen-facing digital services (e.g. http://healthcare.gov or http://recreation.gov) as well as internal digital services and business systems. It is not designed to be used for commodity IT purchases, especially where off-the-shelf solutions can be used as-is at a lower cost to the government.

The "Digital IT Acquisition Professional Training Program," with a curriculum based in principles of agile software design was launched so acquisition professionals could gain valuable hand-on experience applying modern IT procurement strategies. In addition, almost every major buying agency and a number of smaller ones have identified Acquisition Innovation Advocates (AIAs) as primary resources to help the workforce in testing and adopting new, improved, or underused acquisition strategies to reduce delivery times, improve customer and vendor satisfaction, increase access to innovative contractors and lower transaction costs. AIAs seek to achieve these goals by sharing information and tools about contemporary mechanisms to increase the use of effective innovative acquisition practices, streamlining contract processes; and partnering across agencies and with industry..

Challenges, risks and solutions to overcome obstacles

An important lesson learned relates to professionalisation: A policy can be developed, but if skilled and trained staff to implement and the right tools and trainings are missing, the programme will not be implemented effectively and will not meet or achieve the intended results. To overcome these obstacles, United States is hiring resources and/or training internal staff to implement policies for procurement for innovation.

The United States provided the following list of activities that helped to support procurement for innovation and overcome obstacles related to IT services:

1. Consolidate data centres: The original goal was to consolidate at least 800 data centres by the end of 2015. The Office of Management and Budget has since increased the goal to 1200 data centres. Create a data centre "marketplace". The online list allows agencies needing data capacity to find agencies with extra capacity.
2. Cloud-first: Chief Information Officers identified "must move" services and created a plan for migrating those services to the cloud.
3. Contract vehicles for cloud Infrastructure-as-a-Service: The General Services Administration will make available a set of contract vehicles for cloud-based IaaS solutions.
4. Contract vehicles for commodity services: A Software-as-a-Service E-mail Working Group is developing technical requirements for cloud email.
5. Shared services: Agencies develop a "roadmap" for shared services. Programme Management.

6. IT programme management career path: The Office of Personnel Management and the Office of Management and Budget will create an advancement path, including recruiting and hiring, for IT programme managers.
7. Scale IT programme management career path government-wide.
8. Integrated programme teams: IT programmes must include multi-disciplinary teams, including agency leaders and professionals in the IT, acquisition, financial management and legal departments.
9. Collaboration platform: The Federal Chief Information Officer Council will develop an online portal to share best practices.
10. Technology Fellows Programme: Chief Information Officers will recruit by partnering with universities with "well-recognised" technology programmes.
11. IT programme manager mobility: Agencies will offer rotational opportunities to share knowledge and expertise across government. Align acquisitions with technology cycle.
12. Cadre of specialised IT acquisition professional: The Office of Federal Procurement and CIOs will design training for acquisition professionals to develop specialised knowledge to speed up complex IT acquisitions.
13. The Office of Federal Procurement Policy is developing IT acquisition best practices
14. Contracting guidance for modular development: the Office of Federal Procurement Policy will develop templates and samples for modular contracting practices that allows flexibility for evolving technical requirements.
15. Increase opportunities for small tech companies: Align budget process with technology cycle.
16. Work with Congress on flexible IT budget models.
17. Develop supporting materials for flexible IT budget models: The Federal Chief Information Officer Council will create "playbooks" with best practices for this kind of funding.
18. Scale flexible IT budget models more broadly.
19. Consolidate commodity IT spending under the Chief Information Officer: Improve accountability.
20. Strengthen Investment Review Boards: These boards were created to evaluate the results of major IT investments. The plan calls for restructuring the boards to the "TechStat"-model.
21. Redefine role of Chief Information Officer: Chief Information Officers' role will shift from mostly policymaking and infrastructure maintenance to portfolio management. Agencies must terminate at least one-third of their poorly performing projects. "TechStat"-model at bureau level Chief Information Officers and agency leaders will roll out tools and training for "TechStat"-like sessions. Engagement with Industry.
22. "Myth-busters" campaign: Myths that industry and government can't engage with each other creates an "artificial barrier" and reduces agencies' access to market information. The Office of Federal Procurement Policy issued a memo in January

2011 that supported discussions and outreach efforts with key stakeholders. These efforts were expanded with a 2012 "Myth-busters" memo directed to industry partners, and a 2017 edition focused on debriefings for unsuccessful offerors.

23. Interactive platform for pre-RFP (request for proposal): launching a government-wide, online platform for agencies to tap into industry knowledge prior to issuing a request for proposal.

Measurement and impact assessment

The United States monitors the above-mentioned quantified targets related to procurement for innovation. Portfolio statistics sessions serve to monitor agencies' actions regarding the underperforming projects in IT Portfolio, regarding the mandated shift to "Cloud First" policy, and regarding the request to reduce the number of federal data centres by at least 800 by 2015. Additionally, benchmarking metrics measure progress on functional area (e.g. contracting, finance, human capital, information technology, and real property). The quantified target for procurement for innovation is measured; activities to reach the target are followed up by studies of state of play.

Annex B

A template for drafting a strategic action plan for capacity building in public procurement

Task	Institution(s) in charge	Objectives	Beneficiaries	Duration
Organisation of the steering committee	Ministries Procuring entities Universities Training institutions	Conduct the strategy	All stakeholders	1-5 years
Assessment of procurement capacity	Internal service or consultant	Assessment report	Steering committee	3 months
Organisation of the training for professionals Training the Trainers (TtT) and Profs-to-Profs programme	Steering committee +universities and training Institutions under the supervision of the steering committee	1. Define the engagement of the universities and faculties 2. Select the future trainers to be trained in the TtT programme and in the Profs-to-Profs programme	Trainers	3 months
Preparation of training materials	Small group of professors and professionals selected by the steering committee following proposals by the universities and ministries	Creation of modules (documents, presentations) covering national and local public procurement issues + e-learning tools for self-guided learning	The trainers of trainers (TtT) The profs-to-profs programme The trainers and the students	4 months
Piloting the training Accreditation of the training	Steering committee	Verification of the contents Verification of the seriousness of the process for delivering certificate or diploma Collect feedback in order to increase effectiveness	Universities, students, young professionals, professionals	Yearly
Professionalisation: Reform of the civil service legal framework	Steering committee/ o-ordination with the Ministry of Civil Service	Adoption of provisions on the procurement carrier, on the commitment to work for public service and the cool-off period; code of ethics and business manual; incentive for procurement staff	Procurement staff	1 year

Task	Institution(s) in charge	Objectives	Beneficiaries	Duration
Harmonisation and certifications	Steering committee/ co-ordination with the ministry in charge of education	Provisions to regulate the enrolment, graduation, certification of the agreements with training institutions	Trainees and students	1 year to build the certification system
Dissemination of knowledge on public procurement	Steering committee (with consultants) develops materials (e-tools, films, radio messages, pamphlets, workshops for small and medium-sized enterprises [SMEs])	Explain public procurement principles and links with good governance, anti-corruption and public savings	Civil society Non-governmental organisations (NGOs) Media Justice sector Private sector	3 years
Research	Universities + twinning arrangements with foreign universities for students and professor exchange programmes + partnership with Chambers of Commerce for special research and financing of awards	Develop research on public procurement (thesis, PhD research) Create websites and publications (legal journal) on public procurement topics Create clinics in law schools on public procurement issues	Graduate students in law, economics or management	Long term
Monitoring	Steering committee: Report on the results achieved, each year	Performance indicators: Number of trainees/students Results (number of certificates/diplomas delivered) Audience satisfaction surveys Follow-up employment Incentive for trainers/professors	Trainers, trainees, students, young professionals, professionals, all stakeholders	Yearly

Note: The template is a general example for capacity building in procurement and should be adjusted according to the specific requirements of strategic procurement for innovation.

Source: OECD (forthcoming), "Roadmap: How to elaborate a procurement capacity strategy", MENA-OECD Network on Public Procurement.

Annex C

OECD Survey on Strategic Procurement for innovation 2015

OECD SURVEY ON STRATEGIC PROCUREMENT FOR INNOVATION 2015

PART I: Strategic Procurement for innovation (ERAC-Questions)
Dimension of policy making which targets Procurement for innovation

The focus of this part is related to the OECD Recommendation of the Council on Public Procurement (C(2015)2-C/M(2015)4 especially on "Innovation" as one of the strategic secondary policy objectives in Public Procurement in addition to primary procurement objectives.

The following questions are based on the first 6 questions of the ERAC Questionnaire on Procurement for innovation 2014.

***Q1. Please describe briefly the Strategic framework for Procurement for innovation in your country:**

***Q1.1. Do you have a stand-alone Procurement for innovation action plan?**

Please select one

☐Yes

☐No

***Q1.2 Is your Procurement for innovation action plan part of your country's general innovation or procurement strategy?**

Please select one

☐Yes

☐No

Please specify your answer:

***Q2. What is the scope for Procurement for innovation policy in your country?**
Do you use specific definitions to frame Procurement for innovation? (e.g.: European Commission uses PCP and PPI as defined in the rules for participation of the European Framework Programme for Research and Innovation, Horizon 2020).

Please refer to accompanying "Help Notes" for relevant definitions (e.g. SBIR, SMBA, PCP, PPI).

***Q3. Are there any specific actions in your country to support procurement for innovation at the National/Central/Federal level of government?**

Please select one

☐Yes

☐No

Q3.1 In addition, please list some of the initiatives carried out in your country to support Procurement for innovation:

Please also indicate the authority/organisation responsible.

***Q4. Please describe briefly what you consider to be some of the best practices on Procurement for innovation in your country (policies, financial instruments, programmes, stand-alone cases, etc.)?**

Q4.1 In addition, please provide some specific examples of policy instruments used to support Procurement for innovation in your country:

Please select all that apply

	Select all that apply	Please describe shortly these practices:	Please specify the sector of best practice cases:
Policy instruments	☐		
Financial instruments	☐		
Programmes	☐		
Stand-alone cases	☐		
Other, Please specify:	☐		

***Q5. What have been/are the most significant challenges in developing or implementing policies for Procurement for innovation?**

***Q5.1 Have you been able to overcome these obstacles?**

***Q5.2 What were the main lessons learned?**

***Q6. Do you have a system in place to measure the impact of actions related to Procurement for innovation?**

☐Yes

☐No

Q6.1 Please briefly describe:

***Q6.2 Have you quantified a target for Procurement for innovation?**

Please select one

☐Yes

☐No

***Q6.3 Do you measure/follow up moves to reach of this target?**

Please select one

☐Yes

☐No

Q6.4 Has such target actually been achieved?

Please select one

☐Yes
☐No

*Q6.5 Are there any impact assessments, evaluation studies and/or studies of state of play in your country regarding any type of Procurement for innovation?

Please select one

☐Yes
☐No

Q6.6 How do you measure the impact of Procurement for innovation?

Please select all that apply

☐Impact assessments
☐Evaluation studies
☐Studies of state of play
☐Other, please specify:

Q6.7 Please provide any additional information and material regarding the use of evaluation tools to measure the impact of Procurement for innovation in your country (any relevant document can be forwarded to anne.muengersdorff@oecd.org):

Q6.8 Please specify the reason(s) why no system currently exists in your country to evaluate the impact of Procurement for innovation:

Part II: OECD Observatory of Public Sector Innovation (OPSI) in the field of Public Procurement Strategic Procurement for innovation Case / Dimension of procurement for innovation practice

---

*Q7. At what level of government does your organisation operate?

Please refer to accompanying " Help Notes". Please select one

☐Central/Federal
☐Regional/State
☐Local

***Q8. In which sector does your organisation operate?**

Please refer to accompanying "Help Notes". Please select all that apply.

☐General public services
☐Defence
☐Public order and safety
☐Economic affairs
☐Environmental protection
☐Housing and community amenities
☐Health
☐Recreation, culture and religion
☐Education
☐Social protection

Q9. If necessary, please specify your answer:

***Q10. How many people are employed in your organisation (approximately)?**

Q11. Can you provide in national currency the annual budget of your organisation?

Please provide data for 2014 or more recent available year.

***Q12. What is the name of the Procurement for innovation case?**

Q13. If available, please provide a web link that provides further information about the innovation (description, image etc.):

***Q14. What year was the innovation launched?**

***Q15. Please briefly describe the innovative practice:**

In your description please also set out what the innovation has changed compared to the situation before

*Q16. Please check the box(es) which most closely corresponds to the main type of change that your innovation has implemented:

Please refer to accompanying "Help notes". Please select all that apply.

☐It is an entirely new or significantly new service or product, or an existing service provided to a new group of users
☐It is a new process or approach for the design or delivery of existing services or products
☐It is a new way to organise or manage work within your organisation
☐It is a new way to communicate with service users
☐Other, please specify:

Q17. Please indicate the innovation's main beneficiaries:

Please provide an estimated number of beneficiaries where possible.

Q18. Please provide up to 5 key words that capture the key elements of the innovation:

Please provide your answers using the following format: "1. [key word]; 2. [key word]; 3. [key word] etc."

Q19. Why was the innovation developed?

Please refer to accompanying "Help Notes"

Q20. Please set out the objectives of the innovation:

*Q21. Are you aware of existing practices which are similar to your innovation?

Please select one

☐Yes
☐No

*Q21.1 Where did similar practice occur?

Please select all that apply

☐In my own organisation
☐In my public administration
☐In other countries' public administrations
☐In the private sector
☐In the civil society
☐Other, please specify:

Q21.2 Name of existing similar practice:

Q21.3 Organisation where existing similar practice has occurred:

Q21.4 Please provide a description of the existing similar practice:

Q21.5 If possible, please provide a link to any websites describing existing similar practices:

***Q22. Which stage is your innovation currently in?**

Please refer to accompanying "Help Notes". Please select one.

☐Design
☐Testing
☐Implementation
☐Diffusion

Q22.1 How and by whom was the idea for the innovation generated? (e.g. frontline staff, policy planning staff, prize or awards, consulting with users, consulting with others).

Q22.2 Please outline any methods used to test the innovation? (e.g. trialling, piloting, prototyping etc.)

Q22.3 What tools were used to develop and implement the innovation? (e.g. ICTs, management approaches)

Q22.4 Where the information is available, please indicate the resources that were used? (e.g. staff, budget)

Q22.5 Please outline the methods used to extend your innovation to other services or organisations:

Q22.6 What challenges were encountered and what solutions were developed?

Q22.7 Please indicate approximately how much time the innovation took?

Please answer in ***months*** *(e.g. 6; 8; 12; 24 etc.)*

Q22.8 Where you have the information, please also indicate how much time each phase took:

Please provide your answer in ***months***

	Time in months
Design	☐
Testing	☐
Implementation	☐
Diffusion	☐

*Q23. Did you partner with any actors?

Answers may include multiple partnerships. Please select one.

☐Yes
☐No

Q23.1 Partner(s)' sector:

Please select all that apply

☐Private sector
☐Civil society
☐Academics and research bodies
☐Other, please specify:

Q23.2 Partner's name:

Q23.3 Please describe the nature of the partnership and how it affected the innovation:

Q24. What results has the innovation had?

Please check the box indicating the type of results. Please provide evidence of the improvements, including an indication of how they have been measured. Please refer to accompanying "Help Notes".

	Please select all that apply:	Please specify your answer:
Improved efficiency		
Improved effectiveness		
Improved service quality		
Accessibility		
Responsiveness		
Reliability		
Improved user satisfaction		
Results not available yet (please indicate any preliminary results and/or when do you expect to have fuller results available)		

*Q25. Has the innovation been evaluated?

Please select one

□Yes

□No

Q25.1 Please describe the evaluation process used and the results:

Please refer to accompanying "Help Notes"

Q25.2 If possible, please provide a web link to further information on the evaluation:

Q26. What lessons from your experience would you like to share with other countries?

Please include what worked well and less well

Q27. What conditions do you think are necessary for the success of the innovation?

Please refer to accompanying "Help Notes"

Q28. Is there any other information that you would like to share about your innovation?

***Q29. Would you like to make your innovation visible to the public?**

Please select one

☐Yes

☐No

Q30. On a scale of 1 (very difficult) to 10 (very simple) how simple did you find it was to complete the form?

Q31. Were there any questions that you found particularly difficult to answer? If so, please highlight the question and briefly explain why.

ORGANISATION FOR ECONOMIC CO-OPERATION AND DEVELOPMENT

The OECD is a unique forum where governments work together to address the economic, social and environmental challenges of globalisation. The OECD is also at the forefront of efforts to understand and to help governments respond to new developments and concerns, such as corporate governance, the information economy and the challenges of an ageing population. The Organisation provides a setting where governments can compare policy experiences, seek answers to common problems, identify good practice and work to co-ordinate domestic and international policies.

The OECD member countries are: Australia, Austria, Belgium, Canada, Chile, the Czech Republic, Denmark, Estonia, Finland, France, Germany, Greece, Hungary, Iceland, Ireland, Israel, Italy, Japan, Korea, Latvia, Luxembourg, Mexico, the Netherlands, New Zealand, Norway, Poland, Portugal, the Slovak Republic, Slovenia, Spain, Sweden, Switzerland, Turkey, the United Kingdom and the United States. The European Union takes part in the work of the OECD.

OECD Publishing disseminates widely the results of the Organisation's statistics gathering and research on economic, social and environmental issues, as well as the conventions, guidelines and standards agreed by its members.

OECD PUBLISHING, 2, rue André-Pascal, 75775 PARIS CEDEX 16
(42 2016 36 1 P) ISBN 978-92-64-26581-3 – 2017

www.ingramcontent.com/pod-product-compliance
Ingram Content Group UK Ltd.
Pitfield, Milton Keynes, MK11 3LW, UK
UKHW051502290926
15471UKWH00008B/1363

9 789264 265813